Kangaroos, Opossums, and Other Marsupials

Marie M. Jenkins

drawings by Matthew Kalmenoff

HOLIDAY HOUSE · New York

Dedicated to
CARMELITA

Printed in the United States of America

Library of Congress Cataloging in Publication Data

Jenkins, Marie M. 1909-
Kangaroos, oppossums, and other marsupials.

Bibliography: p. 151.
Includes index.
SUMMARY: Examines in general the history and characteristics of marsupials, and discusses specifically the habits and characteristics of opossums, kangaroos, and other similar animals. Includes a section on monotremes.
1. Marsupialia—Juvenile literature. [1. Marsupials. 2. Monotremes]
I. Kalmenoff, Matthew. II. Title.
QL737.M3J46 599'.2 75-10798
ISBN 0-8234-0264-9

Contents

Acknowledgments

I am indebted to Dr. George T. Farmer of the Department of Geology, Madison College, for his assistance in obtaining copies of recent books and articles on moving continents, and I appreciate his willingness to read certain portions of the manuscript and the helpful suggestions he made concerning it. I wish also to express my thanks to Mr. Thomas McLaughlin, Reference Librarian, Madison College Library, and Mrs. Janis Pivarnik, Assistant Reference Librarian, for their aid in obtaining copies of needed scientific papers and the loan of certain books.

Special thanks go also to the following scientists whose letters, comments, and suggestions proved valuable: Mr. Michael Archer, Curator of Mammals, The Queensland Museum, Brisbane, Queensland; Dr. W. D. L. Ride, Director, Western Australia Museum, Perth, Western Australia; Dr. Hobart M. Van Deusen, Curator, Archbold Collections, American Museum of Natural History, New York; Mr. Robert Warneke, Senior Research Officer, Fisheries and Wildlife Division, Melbourne, Victoria; Dr. Patricia Woolley, Department of Zoology, La Trobe University, Melbourne, Victoria; and Dr. H. J. Frith, Chief, Division of Wildlife Research, Canberra, Australia.

1

After the Dinosaurs

Near the edge of the forest a small, opossum-like creature crouched under an overhanging rock. Her beady, black eyes peered through the lower branches of a bush that grew beside the rock and shadowed it. Not far in front of her a fight was taking place.

She watched as two huge beasts faced each other. Tyrannosaurus, the largest and fiercest of all the dinosaurs, was hungry. Reared up on his hind legs and tail, he towered high in the air. His great mouth opened wide as he lunged at another dinosaur.

Triceratops, a heavy, three-horned creature, was not afraid. His head was nearly as long as a man is tall. One stout horn stuck up above his nose and two longer, dagger-like horns jutted out above his eyes. The back of his skull flared out into a bony frill that covered his neck and part of his back. Even the big teeth of the king of dinosaurs could hardly crunch through all that.

The opossum did not fear the two beasts. She was smaller than one tooth of the hungry flesh-eater, and he would not waste his time on such a creature. The three-horned giant ate only plants.

Newborn Infants

Only a few hours earlier the opposum had given birth to seven babies. They were somewhat wormlike and so small

that two or three dozen of them could have been carried in a teaspoon. As each one was born the mother licked it. Then she watched as the infants, one by one, used their front paws to climb through her fur to a pocket, or pouch, on her belly.

These first opossum-like creatures that lived millions of years ago were probably much like our opossums today. The babies are born before they have hardly begun to grow.

No hair is on the skin. Round, dark spots show where the eyes will develop, and the hind legs are just two little bumps. Only the front legs and the mouth are well formed.

Each newborn opossum starts to crawl as soon as it is licked free of the membranes that cover it. No matter which way it starts to move, the dead weight of its body hangs downward. This causes the infant to crawl upward. As soon as it reaches the pouch it tumbles inside.

The mother's nipples are in the pouch. Each mite grabs one and hangs on tightly with its strong tongue. There it will stay, safe and warm, until its hair grows, its eyes open, and it is ready to leave the pouch and learn to live outside.

The Smoking Mountain

The mother opossum stirred in her small corner under the rock. The dinosaurs were still fighting. Behind them, far in the distance, mountain peaks glowed red in the evening sun. From a nearer peak, curls of smoke puffed upward.

The opossum was restless. Strange, faint rumblings reached her ears. These were not forest sounds such as she was used to hearing. Sometimes the ground seemed to quiver under her feet.

Suddenly the evening stillness was shattered. With a deafening roar the whole top of the smoking mountain exploded into the air. Flames belched from the crater. Swirling dark clouds spewed out and settled down over the countryside.

The dinosaur fight was forgotten. Animals fled in terror. The opossum left her sheltering rock and dashed toward the river. Just as she reached its muddy banks the tallest dinosaur caught up with her. One great foot stepped on the small creature and many tons of flesh pushed her down into the mud.

The dinosaur kept going. He had not even felt the slight bump under his foot. He did not know or care that he had crushed the life from one small mother and seven newborn infants.

Many animals besides the opossum lost their lives that night and during the days that followed. Some of the big dinosaurs were not able to escape the sulphur-smelling gases that rolled down from the volcano. They toppled over, and lay where they fell. Smaller dinosaurs and little furry creatures drowned as crowds of animals surged and pushed in the river.

Fossils Are Formed

Later, rains came. Mud from the hillsides washed over the bodies. Floods carried many downstream, some as far as the river mouth. The volcano cooled a little, then it became active again and shot forth fire and death gases. Earthquakes followed.

As the years passed, the world changed. The river became choked with mud and dried up. Mountains were heaved up from the earth. Then they were worn down again, bit by bit, with whipping sand and scouring rain.

The river mud became rock. Softer parts of the animals had rotted away, but bones and teeth remained buried. Most were broken, and some were lost, but many stayed in place and became rocklike, too.

A million years passed. Then another million, and another and another. Finally, a hundred million years had come and gone. During this time many changes took place in the land where the little opossum had lived and died.

Sometimes waters from the ocean swirled over it, making a vast, shallow sea that lasted for a hundred thousand years. Layer after layer of silt and sand were laid down. Later these became rock. In some places the layers of rock were hundreds of meters deep. (One meter is about three and a third feet.)

New rivers cut through the rock and broke off chunks.

Wind and water helped change the pieces back to soil again. New forests grew, and died, and were covered. More rock layers formed. Earthquakes and other earth movements often twisted the layers into odd shapes.

Stories in Rocks

Today in North America, and in many other parts of the world, paleontologists study such rock layers. These are trained scientists who can "read" stories that have lain hidden in rocks for millions of years. Minerals are buried in the layers, as well as fossils, the remains of once-living animals or plants. Many of the minerals and fossils can be used by scientists to find out how old the rocks are.

Certain rock layers are so ancient that nothing that was once alive can now be found in them. They are billions of years old. Some are much like the bits of rock that astronauts brought back from the moon.

The oldest rock layers that contain very many fossils were laid down during a period of time that we call the Cambrian. This name is an old Latin word for Wales. It is used because one of the first studies of Cambrian rocks was made in Wales over a century ago. Many of the animals whose skeletons are in these rocks are relatives of the lobsters and pill bugs that are alive today.

Scientists have given names to all the different kinds of rock layers, according to how old they are. All are grouped into three big sets called eras. The oldest era is known as the Paleozoic. This is a word which means "ancient life." The Paleozoic Era started about 600 million years ago and lasted around 375 million years.

Each era is divided into several periods. Rocks in each

period are different from the others and contain different kinds of fossils. The Cambrian Period is the oldest one in the Paleozoic Era. It lasted about 100 million years.

The Age of Reptiles

The time of the second or middle set of rock layers is called the Mesozoic Era. This word means "middle life." This era is often called the Age of Reptiles because these scaly creatures were the main animals that lived at that time.

The Mesozoic Era lasted about 160 million years. It was during this time that the reptiles we now know as dinosaurs lived on earth. Some were little creatures no larger than a chicken. Others became the largest and most terrifying beasts that ever appeared on land. Several kinds of early reptiles lived in seas or lakes, and could swim. A few groups developed wings and could fly.

The Mesozoic Era is divided into three periods. The third or last one is called the Cretaceous Period. This name came from the Latin word for "chalk" and is used because so many of the fossils have been found in chalky limestone deposits. The Cretaceous Period lasted about 70 million years. The largest dinosaurs lived during this time, but before the period was over all the dinosaurs had died out completely. No one really knows why this happened, although there have been many guesses. The only reptiles left on earth today are such creatures as snakes, lizards, crocodiles, and turtles.

Mammals and Reptiles

The first small mammals also appeared during the Mesozoic Era, probably before the Cretaceous Period began. By the end of the Cretaceous there were several kinds of mammals, but all were quite small. These early mammals developed from mammal-like reptiles. Little by little, as the centuries went by, different parts of the bodies of certain kinds of reptiles changed, and mammal structures appeared.

When animals are alive it is easy to tell the difference between reptiles and mammals. Scales cover a reptile's body, and the skin is dry because there are no glands in it. Young reptiles hatch from large, yolky eggs much like those of birds.

Mammals have hair or fur instead of scales, and many sweat and oil glands in their skin. When the young are born they are fed on milk produced by mammary glands. These glands are the reason for the name "mammals."

It is difficult for almost everyone to tell whether a fossil was once a mammal or a reptile, or some other kind of animal, but a paleontologist can see many differences. For example, a mammal has one bone in the lower jaw and three bones in the middle ear. A reptile has only one bone in its ear, but several in its jaw.

Kinds of Teeth

Differences in teeth are quite important. Turtles are toothless, but teeth in crocodiles and other reptiles are all much alike. They are pointed or peglike, and are used mainly for holding prey so the reptile can swallow it.

Mammals have several kinds of teeth. In front are biting teeth or incisors. We have two pairs in the upper jaws and two matching ones below. Some mammals may have four or five pairs, and others none at all.

Behind the biting teeth are the canines. These are sometimes called "dog teeth" or "eye teeth." We have one canine on each side, above and below, and it is just about the same size as our front teeth. In meat-eating mammals, such as wolves and tigers, canines are extra long. They can be used to stab the prey.

Cheek teeth lie behind the canines. The front cheek teeth are known as premolars and the back ones molars. We use our cheek teeth for chewing. The back ones especially are somewhat flat on top and have little bumps called "cusps." Bumps on lower and upper teeth fit together in such a way that coarse food can be ground between them.

The front cheek teeth in meat-eaters are shaped like shearing blades. They can be used to slice meat off bones. Horses, cows, and other animals that eat plants have large, bumpy grinding surfaces on the cheek teeth. Often there are ridges. These help the animal chew rough leaves or plant stems. Canines are sometimes small in plant-eaters, or they may be missing altogether. This leaves a space between the front and cheek teeth. When a bridle is put on a horse, the bit is placed in this space.

Thousands of fossils of all kinds have been found in Mesozoic rock layers. One enormous dinosaur skeleton reported from Colorado in 1973 was so long that if the animal were alive today it could look in at the top windows of a five-story building. Its teeth are those of a plant-eater. Very small mammals that lived at this time had teeth suited for eating insects, fruits, and small plants.

About the time the dinosaurs died out, mammals began to grow larger and to spread everywhere. Some were fitted or adapted to live in the woods. Others became suited for mountains, plains, or deserts. A few kinds, such as the ancestors of whales and seals, took to the oceans. Some developed the ability to fly and became bats, the only flying mammals.

The Mammals Come on Stage

When the dinosaurs died, the Age of Reptiles was over. A new era, the Age of Mammals, began. This new era is called the Cenozoic, a word which means "recent life." It started about 65 million years ago and is still going on. Nearly all the four-legged animals with hairy covering came into existence during this time.

The Cenozoic Era is usually divided into two periods, called the Tertiary and the Quaternary, but scientists find it more convenient to divide each of these again and form smaller subperiods called epochs. The subperiods are quite important, because it is through a study of fossils produced during these times that we can find out more about our mammal ancestors.

The names of the subperiods are given in the table. Some lasted a much longer time than others, but they can be told apart because rock layers and fossils in each one are different. Some of the subperiods will be mentioned many times in this book. When you need to refer to one, you can use the table to look it up quickly and easily.

Epochs of the Cenozoic Era, the Age of Mammals

Name of Period	*Name of Epoch*	*Meaning of Epoch Name*	*Began (millions of years ago)*	*Lasted (millions of years)*
Quaternary	(RECENT)*		?	10,000+ years
	PLEISTOCENE	most recent life	3	3
Tertiary	PLIOCENE	more recent life	12	9
	MIOCENE	less recent life	27	15
	OLIGOCENE†	still less recent life	37	10
	EOCENE	dawn of recent life	55	18
	PALEOCENE†	oldest recent life	65	10

* Recent discoveries show that early man appeared over a million years ago. Many scientists believe that we are still in the Pleistocene epoch, or "Age of Glaciers."
† These two epochs were added later.

Mammals Advance

Mammal bodies have many improvements that are not found in reptiles. Scales of reptiles, for example, are a good protection against knocks and scratches, or loss of body water, but they are not much use in keeping the animals warm or cool. A snake or a lizard can be killed by the hot sun.

Mammal hair, however, is a good insulation. It traps body heat in winter, and helps keep out hot air in summer.

Oil from skin glands helps make the hairs waterproof, and animals can swim in icy water without becoming chilled.

In animals with sweat glands, the moisture evaporates on hot days and cools the skin. You can see how this works by wetting one hand, then holding both in front of a fan. The wet one becomes cool quickly. Some animals lose water from the lungs by panting. Certain others, such as cats, lick their paws and rub them on the fur, then let the wetness evaporate. When animals lose water they must make up the loss by taking in extra amounts.

Mammals also have brains that are much more developed than those of reptiles. The brain itself is larger, especially that part which has to do with gaining knowledge. Mammals can learn by experience and can teach their young. In fossils the brain cases of mammals are larger than those of reptiles of a similar size and are shaped differently.

Eggs and Infants

Probably the greatest difference between these two great groups is the way they produce and care for their young. Reptiles have large, yolky eggs with much stored food. Usually the eggs are laid where the warm sun can shine on them until they hatch, although a few reptiles carry the eggs in their bodies. In either case the newborn creatures must find their own food and take care of themselves.

Mammal eggs are so small they can hardly be seen with the naked eye. The mother carries the developing young inside her body in a special sac, the uterus, for weeks or months. When they are born she nurses them with milk until they are ready to take other food.

During the time that young mammals stay with their

mother they learn many things. Lion and tiger cubs find out how to stalk prey. Wolf pups are trained to hunt with the pack. Young fawns are taught what dangers to keep away from, and when to hide or to run.

Changes in body structure and in behavior went on in the different groups of mammals for millions of years. Some kinds could not adapt very well. They lived for only a few million years, then died out, or became extinct. Other groups still live today. A few are called "living fossils" because they are so much like their ancestors that lived millions of years ago. Most species today are not exactly the same as the ancient ones because some kind of change is always going on.

Placental Mammals

Long before the Cretaceous Period ended the mammals had divided to form two great groups. One of these is the placental mammals. Most of the four-legged animals that we know belong here. Bears and bats, dogs and cats, and monkeys and elephants are all placentals. Human beings are included in this group too.

In all kinds of animals, when development first begins, the growing infant is called an embryo. During this stage all the organs and systems of the body are begun. As soon as the beginnings of its organs are formed, the embryo is called a fetus. The rest of the time it is in the uterus it is growing larger, and is becoming the same particular kind of animal that its parents are.

In placental mammals a special structure called a placenta develops on the wall of the uterus. It is made partly from membranes that surround and protect the embryo, and partly from the inside of the uterus wall. As the embryo grows, the placenta grows also. Food and oxygen from the mother's bloodstream go through the placenta, then through a long stalk that reaches to the developing infant. This stalk is the umbilical cord, sometimes called a navel cord. When a placental mammal is born, the mother bites off the cord. The stump soon shrivels and forms a small scar, the navel.

Soon after birth the uterus throws off the placenta also. It is sometimes called the "afterbirth." In human babies it is about the size of three or four breakfast plates stacked together. Doctors or nurses in hospitals cut the cord and throw the placenta away.

Marsupial Mammals

The other branch of early mammals is the marsupials. Most of these animals have a marsupium, or pouch, on the belly where the young are carried after birth. No placenta grows in marsupial mammals, so the young are still embryos when they are born. They finish development in the mother's pouch while gaining nourishment through her nipples. In a few species there is, for all practical purposes, no pouch at all.

Newborn marsupials may stay in the pouch for weeks or months. At first they leave only for short periods, then run back and dive into the mother's pouch again. When they become too large to crawl inside, they must learn to live alone.

Fossil bones and teeth of marsupials are different from those of placentals. Pouched animals have a pair of marsupial bones growing forward from the front of the bony girdle about the hips. Some scientists think these may be of help to the mother when she carries heavy young ones. This must not be the only use, however, for male marsupials have these bones also. No such bones are ever found in placentals.

Marsupials have more biting teeth than placentals do. Also, at the back of each lower jaw is a small, angle-shaped piece that is turned inward. No other kind of mammal has this structure.

Pouched mammals are not common over the world. Most families of marsupials live in Australia, but only a few kinds of placentals are "native" there. South America has several different kinds of marsupials, as well as many marsupial fossils.

In all of North America, however, only one kind of marsupial is living today. This is the opossum, a silvery-gray creature about the size of a large cat. It is our most unusual mammal.

2

Our Only Pouched Animal

The Virginia opossum of North America was a source of wonder to early settlers. They had never seen a creature that bore babies so small they had to be carried in a pocket on the mother's belly until they grew large enough to climb outside and cling to her fur. All sorts of legends grew about them.

Some thought the mother breathed her young out through her nose. Probably this idea developed because people had seen the animal nosing and licking her pouch before the infants were born. Others believed the babies grew out from the nipples, much as apples grow from buds on apple trees. How else, they wondered, could such bean-sized midgets cling so tightly that they seemed to be glued in place? Four hundred years passed before the mystery was settled and people knew how young opossums really are born.

Captain John Smith, whose life is said to have been saved by an Indian princess, Pocahontas, first gave the name "opossum" to the animal. The word comes from an Indian term that means "white face." When Captain Smith lived, opossums were found mainly in southern states, but they have now spread to the west and north. A few even live in Canada.

Animals with "Five Hands"

An adult opossum has a long, pointed face with a pink nose and round, shiny, black eyes. The ears are thin and hairless and can be folded over the ear openings when the animal sleeps. The fur is rough, with dark-tipped guard hairs (long, coarse hairs rising above the undercoat) scattered through it.

An opossum's paws are dark, but the toes are pink and are spread apart as if they were fingers. When the animal walks, its feet leave starlike tracks. The first toe on each hind foot has no claw and is "opposable." This means it can be used as a thumb. An opossum holds food or tree limbs with its paws in the same way that you or I would grasp them with our hands.

The long, scaly tail of an opossum is sometimes called its "fifth hand." Twined around a branch, the tail helps balance the animal and leaves the front feet free. Young opossums sometimes hang by their tails, but grown ones are usually too heavy to do this.

An opossum has 50 teeth. This is more than any other mammal in North America has. Five pairs of biting teeth are in the upper jaw and four pairs in the lower. The canines are long, and the cheek teeth have cusps, or bumps, that help grind food. Teeth like these are found in omnivorous animals, those that eat all kinds of food. A good meal for an opossum might include anything from snails to persimmons, garbage to grasshoppers, or mice to lizards.

Opossums live in woody or swampy areas near streams. Each lives alone, although several may share the same territory. They rove about, hunting food. If it is scarce they may wander great distances.

Summer and Winter Behavior

Like other mammals, an opossum is "warm blooded." This means it can control its body warmth so that it stays at about the same temperature whether the outside air is cold or hot. A marsupial's body temperature is usually a little lower than that of a placental mammal the same size.

Weather extremes are hard on opossums. On hot summer days when the temperature reaches much above 28° C. (over 80° F.) the opossum cools its body by licking it or by licking its paws and rubbing them over its fur. If water is nearby, it may jump in and take a swim.

A scientist who tested swimming abilities in small mammals found that an opossum could stay in the water several hours at a time. Mother opossums carrying babies can swim, too. A muscle around the opening of the pouch can be puckered so tightly that no water can enter. You use a muscle much like this when you close your lips tightly.

When the pouch is closed, the infants inside must breathe the same air over and over again. As a result, much carbon dioxide is formed. This is a waste gas given off by all animals that take in oxygen.

Most animals are killed by a large amount of carbon dioxide, but experiments have shown that young pouch opossums are not harmed. They can stand a much greater amount of carbon dioxide than weaned ones can. In this way they are like unborn placental infants. Bodies of such young animals seem not to have developed yet to the point where a great deal of oxygen is necessary.

In the North, when the temperature dips below freezing and stays there for several days, woodchucks and many other small animals are not harmed. They store food in their bodies, then hibernate, or become torpid with the slowing-down of their body processes, for several weeks during cold weather, and live on the fat. An opossum cannot do this.

When a storm comes it must find shelter and hide. If

no other warm place is near it may even share a den with a raccoon or a skunk, and sleep for a few days. Then it gets hungry and must go out in the cold to hunt food. As a result, opossums that live in the North often have frostbitten ears or tail tips.

"Playing Possum"

Opossums have few enemies. A hungry bobcat, coyote, or hawk may at times take an opossum, but most wild animals seem not to pay much attention to this slow-moving creature. Some years ago Eleanor Dater of Ramsey, New Jersey, kept a feeder outside her window where she could watch wild animals come to eat. Two gray foxes came for food, a raccoon with kits, and some skunks and opossums.

A fox would grab some food then run away to eat. An opossum would take time to look things over, then settle down and help itself. Often a raccoon and an opossum or two ate from the same dish. When a skunk wandered up it might be either snapped at or allowed to eat with the others. The opossums did not disturb the other animals and were not bothered by them.

Opossums are sometimes trapped in the south and their pelts sold. Also, they are hunted for food, with the aid of dogs. An opossum rarely fights. It escapes into a tree if possible, but a cornered animal will turn and face its attacker. It hisses and growls, and threatens with a mouth so wide open one can almost count all 50 teeth. If the bluff fails and the dog grabs an opossum, it may "play dead."

Many animals seem to fake death when attacked, but the opossum is so good at this that the response is called "playing possum." The animal suddenly appears to give up

An opossum "playing dead" can be very convincing.

and slumps to the ground, lying on its side. Its mouth is half open and the tongue hangs out loosely. Sometimes the eyes have a glassy stare.

An opossum in this state can be pushed about, shaken, kicked, or even carried off, without giving any sign that it is alive. When left alone it will "recover" slowly. First it raises an ear, then the head is slowly lifted and the opossum looks around. If nothing happens, it carefully gets to its feet and walks off.

For years people thought the opossum just pretended to be dead. Then scientists began to study the reaction and decided perhaps the opossum became so scared it was paralyzed for a short time. A person who "freezes" in front of a rushing train shows fright in much the same way.

Later studies have shown this probably is not true. Three scientists in California tested a group of opossums with an electroencephalograph, or brain-wave machine. First they made tracings when the animals were awake, then asleep, to find out what their normal brain waves—small electric currents in the organ—were like. Then they tried to scare the opossums with a record of barking dogs, but the animals paid little attention. Next they shook the animals with a big pair of pliers. Finally they tried tests with a real dog.

When they became excited the opossums would hiss and growl, but they did not usually give up and fall over unless the danger was real. Even then, the brain waves were almost the same as normal ones. This showed that animals pretending to be dead were really wide awake and alert. The opossum can control, too, how long it stays in this condition. When the danger seems to be over, whether it is two minutes later or half an hour, the animal "comes to," then gets up and shambles off.

This kind of behavior is a good defense. A dog, for example, will chase a creature that seems to be running away, but it is likely just to nose a dead one for a few minutes. Then the dog goes off, leaving the animal alone, and hunts for something more interesting to chase.

The tail of an opossum is helpful for carrying nesting materials, and in the case of young ones, for hanging from branches.

Nests in the Springtime

A mother opossum usually has her first litter early in the spring. Sometimes she makes a new nest, but she is just as likely to use a den some other animal has made, or to find a hole in a hollow tree. The nest is lined with dry leaves or grass that make a soft bed.

The opossum's "fifth hand" is very useful here. She tucks her looped tail forward between her hind legs, picks up several leaves with her mouth, then uses her forepaws to slide them backward under her body. The hind feet grab the leaves and slide them into the tail loop while the opossum noses around for another mouthful.

Two or three minutes later the loop is filled with six or eight mouthfuls of leaves. Then the opossum holds her tail out behind her, takes a few more leaves in her mouth, and scuttles away. Only a few such trips are needed.

Opossum Embryos

At mating time up to 24 eggs are produced and started on their journey to the mother's uterus. Each one looks much like a chicken egg without a shell, but it is so small that 26 eggs in a row would make a line only as long as a 25-cent-piece is wide. The eggs are fertilized by sperm cells from the father opossum, then development begins inside the uterus.

An opossum embryo is ready to be born when it is not quite 13 days old. By this time it is almost as large as a honeybee and weighs about as much as one or two postage stamps. If a human baby were this much smaller than its mother when it was born, it would take between 600 and 700 babies to weigh as much as an ordinary baby really does weigh.

Usually 13 nipples are in a mother opossum's pouch. The opossum family is the only one in which the nipples are not in pairs, but are arranged in a sort of circle. While the babies are developing in the mother's body, the milk glands grow larger, and the inside of the pouch becomes moist and oily.

Seven days after the egg starts to develop, the opossum embryo is about the size of a pinhead. As it grows the infant receives some nourishment from a yolk sac that has formed from the yellow part of the egg. When this is used up a little food material oozes out from the walls of the uterus for a few more days. No special organ develops which could help transfer food from the mother to the embryo. Soon the young embryos must leave the uterus or they will starve.

Birth and into the Pouch

Shortly before the babies are born the mother becomes restless. She often licks the inside of the pouch and the fur around it, perhaps because the swelling milk glands make her uncomfortable. Years ago people thought the mother was licking a path in her fur for the young ones to travel.

Just before the first baby arrives the mother "sits down" with her tail stretched forward between her hind legs. As each embryo is born it rests on the base of her tail, and the mother leans over and licks it. This breaks and tears away any membranes that cling to it.

Each naked, pink infant then crawls upward toward the pouch and tumbles inside. The mother may watch the infants but she never tries to help them. Each newborn noses about, twisting its head until it has found a nipple about the size of a pinhead. It grabs this and hangs on tightly with its powerful tongue. A young opossum begins sucking as soon as its lips close over the nipple. It uses both its tongue and cheek muscles. If one listens closely to new infants in a pouch he can hear the sounds they make.

Within an hour the teat has been stretched by the squirming youngster until it is twice as long. As the small animal grows and moves about, the teat to which it is attached becomes long and slender. It acts not only as a tube for milk to pass through, but also as a sort of tether or "lifeline" to hold the young animal to the mother.

Within a few days the end of the nipple has enlarged until it fills the infant's mouth. Years ago it was thought that the tissue of the nipple and mouth actually fused, but Dr. Carl G. Hartman of the University of Illinois found that the young ones could be removed without harm if they were handled gently. When put back in the pouch, each small creature would again attach to the first nipple it could find.

A young opossum does not have to let go the nipple to

When a young opossum is born, it is actually smaller than a honeybee. At the right it is seen nursing in the pouch after two days of development there.

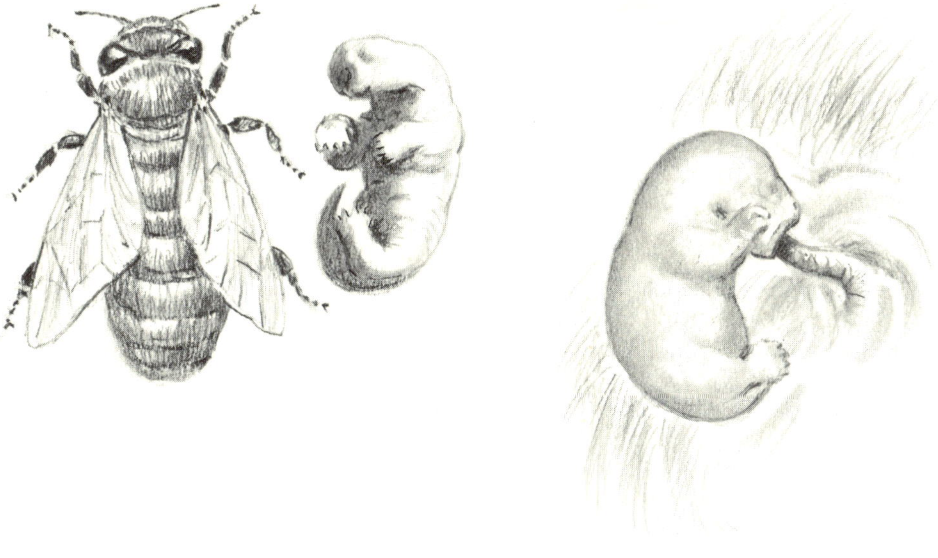

breathe, for its throat is not like that of a human baby. When either a human child or adult swallows, a small "lid" at the back of the tongue flips down over the windpipe and keeps food or water from getting into it. Sometimes, especially if we are laughing or excited, the nerve signals become mixed and the lid does not flip soon enough. Then food or water goes down the wrong tube and we cough or strangle.

This does not happen to the young opossum. Instead of a clear-cut lid it has an air tube that reaches up through its throat to the back of the nose. There is nothing to flip, so the baby can breathe and suck at the same time. As the young opossum grows, this tube is changed into a lid that is much like ours.

For many years scientists thought that marsupial mothers somehow pumped milk into the young animal's stomach, but Dr. Robert K. Enders of Swarthmore College has found that this idea is not true. Dr. Enders found there were no muscles at all connected with the nipples that could cause them to squeeze out milk. He stimulated both nerves and muscles in many ways, but the nipples would release milk only when they were "milked," or squeezed gently on the outside.

Young opossums stay attached to the nipples for about two months while their organs develop. If a six-weeks-old infant is taken from the pouch it can take a step or two, and give some high-pitched cries that sound like sneezes. In another two weeks the infant can curl its tail around a stick and support its own weight. At this age the youngsters sometimes dangle part way out of the pouch, still hanging to a nipple inside.

Nine-week-old opossums are about the size of mice. A week later they are as big as small rats, and begin to run

Young opossums ride on the mother after about ten weeks of so of growth.

about. They climb in and out of the pouch and soon take rides on the mother's back or sides, clinging wherever their small paws can grab a fistful of fur. Their tails often serve as an extra anchor, wrapped about the mother's ears or nose, or perhaps a leg of a brother or sister close by.

The World Outside the Pouch

Young opossums are weaned when they are thirteen or fourteen weeks old. They can now find their own food and will soon be able to live alone, but they wait until the next year before starting families of their own. Most opossums, unless killed in some accident, live to be two or three years old.

The mother wastes no time in beginning a second litter. Often a new brood is safe in her pouch within two weeks after the first one has gone. In the South, three litters a year are fairly common.

Although a mother produces up to 24 eggs each time, not all the infants live. A few never reach the pouch. Some finish the journey but find no nipples left, and are doomed to die. Others are weak and are pushed aside or lose their grip on a nipple. Usually six or eight live until weaning time.

This is one way in which nature is kept in balance. The strongest survive, but those that are sickly or feeble die before they can pass their weaknesses on to their own young. The group as a whole then stays tough and sturdy.

Curious—or Stupid?

The opossum is an unusual animal for more than one reason. It has a reputation for being stupid and slow-witted, but much of this seems to be due to its not being afraid of most creatures. It has much curiosity, and often gets into things it should leave alone. It is more likely to stare at some intruder than to run away.

The brain of an opossum is only about one-fifth the size of a cat's, but some areas are well developed. The part that controls the sense of smell is large, as is the region concerned with hearing. The animal can be startled by a soft camera click, yet a strong light flashed in its face will not bother it a great deal.

The opossum is one of the few animals that has lived on earth almost unchanged for millions of years. Many mammals, such as the giant cave bear or the huge, hairy mammoth, appeared long after the first little opossum-like creature. They faded into history thousands of years ago, but the opossum lives on.

Probably one reason is the animal's ability to adapt and fit in with whatever conditions it finds. It will eat any kind of food, and can live in hot lands or cold. It recovers from broken bones and other injuries that would kill most creatures. Also, one opossum produces more young each year than almost any other animal its size. Since the strongest ones live, the animal's good qualities are passed on to its offspring.

Opossums Help Scientists

Because opossums are so different from our other common animals, they are often used in special experiments. Dr. Merle Mizell of Tulane University has found that young opossum hind limbs can grow again when they are cut off, provided that nerve tissue is supplied to them. Together with Dr. D. A. T. New, of the University of Cambridge, Dr. Mizell has been able to grow unborn opossum embryos in culture dishes until it is nearly time for birth. When more is learned about experiments of this kind, the new discoveries may be a big help to doctors and physicians who treat human mothers.

Scientists have found that many mammals, including the opossum, go through periods of REM, or rapid-eye-movement, sleep. During this time the animal's heartbeat and breathing are nearly the same as when it is awake. An animal in REM sleep wakes quickly and is not likely to be surprised by an enemy.

The opossum spends about one-third of its sleeping time in the REM stage. This kind of sleep has never been found in reptiles. Perhaps the ability to sleep lightly and wake quickly is one reason why mammals have spread over the earth, while many of the reptiles have vanished.

True Opossums

North America's strange little marsupial belongs to the true opossum family. These animals are called "didelphids," a name which means "two wombs." The scientist who named the animals over two centuries ago thought the pouch was a second, outside womb. It was discovered later that most marsupials not only have a pouch but that the uterus on the inside of the mother is double—that is, there are two separate ones.

True opossums are found only in North and South America. We know from fossils found in our western states that several species lived in North America during the Cretaceous Period. Opossums lived here during the first part of the Cenozoic Era, too, but they seem to have died out during Miocene times. No marsupial fossils are known in North America for the next twenty million years. That length of time covers the rest of the Miocene, all the Pliocene, and the first part of the Pleistocene subperiods.

The oldest marsupial fossils known in South America were found in rocks of the Paleocene subperiod, early in the Cenozoic Era. Scientists think perhaps the first marsupials reached South America by wandering down through what its now Mexico and Central America. This long journey probably took millions of years. Once the marsupials reached South America, they spread to all parts of the continent. Their fossils have been found in rock layers of every subperiod of the Cenozoic Era.

Later some of the didelphids seem to have moved northward again. One species, the large common opossum *(Didelphis marsupialis)*, came so far north that it reached the region where our southern states now lie. It has lived here for thousands of years. During this time it has changed in several small ways, and has become a race or variety of the common South American opossum. We now call it the North American or Virginia opossum, and its scientific name shows this change. It is *Didelphis marsupialis virginiana.*

3

Varied Southern Neighbors

Not long after Columbus sailed home from the new land he had discovered, one of his ship captains led a trip of his own. Pinzón landed in what is now Brazil, in South America. While he and his men were exploring one day they found an opossum at the base of an enormous tree. It was a mother, with young ones in her pouch. The men were astonished, for none of them had ever seen such a strange creature before.

When Pinzón went back to Spain he took an opossum with him; probably not a living one, however. At court he presented many treasures to King Ferdinand and Queen Isabella, and the opossum was one of his gifts. The King and Queen were as amazed as Pinzón had been.

No one knows just what kind of opossum Pinzón found, for many kinds of didelphids live in South America. Probably it was the cat-sized common opossum *(Didelphis marsupialis)*, if we judge from descriptions written of it afterward. This opossum ranges through Mexico, Central America, and much of South America. Another species closely related to it is slightly smaller and darker, and is found as far south as Argentina.

Opossum Beginnings

The oldest marsupial fossils known are those of the didelphids that lived in North America at the same time as the great dinosaurs. For this reason most scientists are convinced that all marsupials had their beginnings here, and spread from North America to other parts of the world. South America was probably the first continent reached.

Scientists who study rocks and soils, and rivers and seas, think that a journey overland into South America would have been possible during the time these small marsupials lived. Many facts have been discovered showing that the two great American continents were connected during the Cretaceous Period and for the first part of the Paleocene subperiod. There is also much evidence that this link was later broken, probably before the Eocene subperiod began. From then until the Pliocene, South America seems to have been a great island continent. No animals could reach it from the north for 50 or 60 million years.

During part of this long period of time a great sea washed between the two continents. Sometimes part of North America was flooded as far north as what is now the Ohio River. Later a string of volcanoes appeared in this sea. Some of them became islands. Long afterward more land rose from the sea and became what is now Central America. Finally a neck of land formed a bridge, or isthmus, that connected the northern and southern continents.

This strip of land was only a few miles wide. Much of it was either jungles or mountains. Some animals could travel through this difficult country, but many others could not. For this reason the Isthmus of Panama was known as a filter bridge.

During the millions of years that South America was separated from other continents, many kinds of unusual mammals developed in it. Some have been found nowhere else in the world. One of these was a long-legged, three-toed creature that looked like a cross between a camel without a hump and an elephant with only half a trunk. These strange animals are known only from fossils, for they disappeared from the earth millions of years ago.

The Mouse Opossums

Didelphids that wandered down from North America found all sorts of places to live. The climate was warm, the land rich, and food was plentiful. As millions of years went by the small animals grew and changed, and became suited to live in all kinds of habitats. Mountains and plains, rivers and trees, and even dry deserts became the home of different marsupials. Many new species appeared.

Several members of the true opossum family still live in South America today. Next to the common opossum, probably the most familiar ones are small, rat-sized creatures called mouse opossums. Close to 50 different species are known. These animals usually live in tropical forests, or along the edges of fruit plantations.

Mouse opossums have a smooth, shiny fur that ranges from yellowish gray to dark chocolate. The most common kind *(Marmosa murina)* is a golden brown color above and white below. Its eyes are large and are often surrounded by dark circles. The ears are delicate and naked and can be folded down when the animal sleeps.

Like most of their relatives, mouse opossums live alone. They hide and sleep during the day, and feed at night. Both insects and fruit are eaten, but bananas seem to be a particular favorite. A mouse opossum sits up to eat. It holds the food with its front paws and takes little nibbles with its needle-sharp teeth.

The mouse opossum of South America

Some species of mouse opossums have no pouches at all, and others have only small folds of skin. Newborn embryos are about the size of rice grains. Each one climbs upward and grabs a nipple. For the next few weeks the babies hang in a cluster near the mother's hind legs.

As she scrambles about in the trees the young dangle in all directions. Sometimes one is bounced against a twig and knocked off. In such a case it opens its mouth as if it were screaming, but the sound is too high for human ears to hear. The mother apparently can hear it, however; if such a loss occurs, she turns back to recover the missing infant.

By the time young mouse opossums are a little over a month old their eyes are opened, hair covers their bodies, and they begin to crawl about. Soon they ride on the mother's side or back, gripping her fur with their small paws. A month later they are fully weaned and are able to care for themselves.

"Four-Eyed" Didelphids

Gray four-eyed opossums (*Philander opossum*) are found from Mexico to Argentina. They are not quite as large as the common opossum, and their fur is short and straight. Above each eye is a large white spot. This gives the opossums the appearance of having two black and two white eyes.

These animals live in forests but can swim almost as easily as they climb. They eat all kinds of small creatures, including earthworms, frogs, snails, and even small birds and mammals. Sometimes they raid fruit crops or cornfields.

The gray four-eyed opossum does not "play possum." It chatters loudly when disturbed, but if danger threatens, it opens its mouth widely and hisses. It will fight savagely, using its sharp teeth. Some years ago Dr. Don Wilson, of the University of New Mexico, watched a fight between a four-eyed opossum and a large common opossum. The two animals were on a lawn in the Panama Canal Zone.

At first the animals circled about each other, snapping and hissing. Then each one tried to grab the other's nose.

Finally, after two or three trials, the bigger opossum threw the smaller four-eyed one and killed it by biting it on the head. Then he ate it. The animals were so wrapped up in the battle that they were not stopped by a rain shower, nor by headlights glaring on them.

The pouch in gray four-eyed opossums opens toward the front. Five to nine nipples are inside. Like many other creatures that live in the tropics, these animals may have two or three litters a year.

Brown four-eyed opossums *(Metachirus nudicaudatus)* are a little smaller and darker than the gray ones. The tail is so long and naked that the animals are often called "rat-tailed opossums." There is no regular pouch, but two folds of skin on the belly give some protection to newborn infants. Older youngsters are jiggled about, much as mouse opossum babies are, as the mother jumps from branch to branch in the trees.

Yapoks, the Water Babies

The water opossum, or yapok *(Chironectes minimus)*, is the only marsupial that is more suited to live in water than on land. Its fine, thick fur is waterproof, and the hind feet are webbed so the animal can almost "run" in the water. The thick, naked tail is flat and is longer than the body. This is the only didelphid that does not use its tail as a fifth hand. Water opossums live near mountain streams and lakes from southern Mexico to northern Argentina.

Yapoks nose in the mud for food, collecting clams, crayfish, or water insects. Long, coarse bristles on each side of the face probably help the animal find its prey. The opossum crawls out on the bank to eat, sitting on its haunches and

The yapok is a water-lover.

holding the food with its forepaws. Then it dives into the water again.

A mother yapok often digs a burrow just above water level, with a nest at the end. Older offspring are left here at times, but young ones stay attached to the nipples while the mother hunts for food. The pouch opens toward the rear and a strong muscle closes it tightly, so the infants are safe while the mother swims or noses in the mud of the stream. Five to seven young are born at one time.

Unusual Opossums

Three South American opossums have been shown by blood tests to be different enough from other didelphids that they are often placed in a separate subfamily. One of these, the woolly opossum *(Caluromys lanatus)*, is found in forested areas of northern and central South America. It lives in tree hollows or limbs, and is sometimes active during the day. Like other opossums, it eats almost anything.

The soft, thick fur is a rich golden brown. It covers

The woolly opossum was formerly trapped for its fur, but it is safer from such trapping today.

the body and about half the tail. A dark stripe on the face runs almost to the nose. The animals were trapped for their fur some years ago, but there is little market for the pelts today.

Woolly opossums are unusual in that they are almost as sensitive to red light as to ordinary light. This is not true of most night-loving creatures. For this reason, zoos that show the animals in their natural surroundings have to be especially careful about lighting arrangements.

An Ancient Family

One other small family of marsupials is found only in South America. These are the caenolestids. The name means "recent robbers" or "new pirates," but these mouse-sized creatures are more commonly known as rat opossums. Only three species are known.

A scientific paper on rat-opossum fossils was published over a hundred years ago, but the first living animal in this family was not seen until 1911. Most of the rat opossums live in the high rain forests or grassy valleys of the Andes mountains, although a few are found on some of the cool, mountainous islands of southern Chile.

The small animals are dark-brown, with a soft, dense fur. The head is long and cone-shaped, and the ears rounded. Part of the tail is furry, but the rest is rather four-sided and covered with scaly rings and a few stiff hairs. It cannot be used as a "fifth hand."

The common rat opossum *(Caenolestes obscurus)* lives on grassy, swampy land. Not much is known about how these creatures mate, or brood and care for their young. They have no pouch, but two pairs of nipples lie between the mother's thighs. Probably the infants hang onto nipples tightly, as mouse opossums do, and are bounced around as the mother seeks food.

Fossil Beasts of Prey

Although no fossils of marsupials have been found in Cretaceous rock layers in South America, Paleocene rocks are full of rich deposits. Bones or teeth of both marsupial and placental mammals are found there.

Some of the marsupial fossils belong to a great family that is found only in South America. The animals developed, lived several million years, then died out. They are known as borhyaenids. The name means "greedy hyaena." The true hyaenas of today, found in Africa and southwest Asia, are doglike animals that have many characteristics of cats, but they are not at all related to these long-ago marsupials.

Borhyaenids were meat-eaters. The first ones were rather small, but by Eocene times several species were the size of bears. They had large, strong canines, and their cheek teeth were sharp and able to shear meat from bone easily.

Perhaps the main reason that borhyaenids developed so rapidly and became such an important family is that there were no other kinds of flesh-eaters in all of South America. Before the great sea washed between the two American continents, and kept them separated for millions of years, other mammals as well as didelphids migrated to the south, but all of them ate only plants or insects.

During the time South America was an island continent, many kinds of placental mammals developed, but none of them preyed on other animals. Some were queer hoofed creatures that looked nothing like the horses or cattle or deer that we know today. Probably one of the oddest was a huge beast that ate plants and seemed to be a cross between a rhinoceros and an elephant. It lived only a few million years. Horny-plated armadillos, much like those in our southwestern states, also appeared during this time.

Many kinds of borhyaenids evolved. They had no rivals, for none of the other animals was a carnivore, or meat-eater. Some of the borhyaenids resembled wolves; others suggested weasels or martens. One of the larger ones was a cougar-like animal that scientists today call Borhyaena. It had claws; the back and legs were strong; and the skull was much like that of a large dog with big canine teeth.

Saber-Toothed Borhyaenids

Probably the most unusual borhyaenid was a leopard-sized animal known as Thylacosmilus. It was a marsupial but it looked much like the fierce saber-toothed tigers that roamed the forests of Europe and North America millions of years ago. Thylacosmilus had no front teeth, but the two top canines were long and curved and used for stabbing prey. They fitted into sheaths in the lower jaw. In order for the canines to work well the animal had to be able to open its mouth widely. Fossil bones show that vertebrae in the neck had become so modified that the jaws could spread far apart.

Thylacosmilus was one of the last of the borhyaenids. The family had started during Paleocene times and had developed for 50 or 60 million years. About 10 to 12 million years ago the animals began to disappear. Before the Pleistocene subperiod began, about three million years ago, all were gone.

The Filter Bridge

Extinction of the borhyaenids began about the same time that the Panama filter bridge was formed. Only a few animals from South America moved northward over it. The two best known are the Virginia opossum and the armadillo.

Many kinds of placental mammals came into South America from the north. Century after century rabbits, squirrels, mice, and raccoons clambered over the mountains or wound their way through the jungles. Large and small cats, and wild dogs and wolves made the crossing. Hoofed animals, including horses, llamas, and deer, joined the stream.

These newcomers were rugged and tough. As they moved farther southward, generation after generation, weaker animals died. Only the stronger ones, and their descendants, pushed onward. The animals that finally reached the southern continent, hundreds or thousands of years later, spread into every kind of place.

Many of the South American creatures were no match for these new arrivals. The borhyaenids seemed to suffer the most. Perhaps they were not fast enough to chase and capture the swift-hoofed animals from the north. Or possibly the marsupial carnivores lost out to their placental cousins, the cats and wolves and weasels, because they were not as cunning, or could not creep up as silently and pounce on and kill the prey.

We are not sure just what happened, but we do know that during late Pliocene times the last borhyaenids died. It is only through their fossils that we realize they ever lived.

The didelphids did not die out. Like the small opossum-like animals that outlived the dinosaurs, they could live anywhere, and eat any kind of food. Many young were born each year. Some died, but others lived.

On Other Continents

Except for the Virginia opossum in North America, and the didelphids and caenolestids in South America, all other marsupials now live in Australia or on nearby islands. Neither Africa nor Asia seems to have any marsupial fossils, although rich beds of placental bones and teeth have been discovered on both continents.

Europe once had a few opossums about 15 or 20 million years ago. Georges Cuvier, a great French scientist, discovered the first marsupial fossil there in 1820. It was embedded in a large piece of gypsum rock picked up from the streets of Paris.

Fossil teeth had been found in the rock, but they were not like teeth of any other animal in Europe. Cuvier knew fossils well enough to guess what kind of teeth these were. He took the rock to one of his classes in the Collège de France and announced he would chip away the stone and show the bones of an opossum.

The students laughed. Who had ever heard of pouched animals living in Europe? Nevertheless they watched closely as Cuvier carefully separated bones from rock. When the hip girdle was reached, there were the paired pouch bones such as are found in every marsupial, but in no placentals. Cuvier was right. His students stood up and cheered.

This French scientist is also famous as the first person to show that animals with hoofs and ridged cheek teeth are plant-eaters, while beasts of prey have long claws and canines, and sharp, shearing teeth.

Finally we must ask ourselves this question: If the tiny, opossum-like creatures that lived in North America during the Cretaceous Period were the ancestors of all other marsupials, how did the animals spread to other continents far away? Scientists have argued about this question almost as long as pouched animals have been known. A few answers are now beginning to appear. And it is the geologists—the scientists who study rocks and oceans—and the paleontologists—those who study ancient life—who are providing us with much of the information.

4

Moving Continents and Wandering Animals

Two hundred million years ago our world was very different. All of today's continents were gathered together into one great land mass, and the rest of the earth was covered by a vast ocean. This supercontinent has been named Pangaea, which means "all lands."

Early in the Age of Reptiles, Pangaea began to split apart. A great arm of the ocean slowly pushed inward from the east, separating the continents we know today as North America, Europe, and Asia, from another land mass to the south. At one time these northern continents were in such a different position that the equator ran through the place where New York City now stands.

The first maps of shores on both sides of the Atlantic Ocean were made about four centuries ago. People soon noticed that the outlines of South America and Africa looked as if the two continents might fit together. Not much thought was given to this idea, however, until nineteenth-century geologists began to study rock layers in the southern hemisphere.

Broken Layers

Before many years had passed, some amazing facts were discovered. Certain layers of rock in Africa were so much like others in South America that it looked as if the two continents had broken apart. Many fossils of plants and animals on both sides of the ocean were found to belong to the same species. This would not be possible unless the lands were once joined.

The great land mass that was separated from the northern continents when Pangaea was split in two is known as Gondwanaland. The name comes from an ancient people, the Gonds, who had a kingdom in India long ago. This name was chosen because rock layers just like those in the land of the Gonds had been found in Madagascar and southern Africa, as well as India.

Later discoveries have shown that the same rock layers also occur in South America, Australia, and even Antarctica. In each continent these layers contain thick coal seams that have an enormous amount of fossil Glossopteris leaves. This is an ancient plant that seems to be related both to ferns and to seed plants.

A New Theory

Earlier scientists had thought that the likenesses of rocks and fossils on two sides of an ocean were due to great land bridges that once connected the continents, then sank out of sight. It was said this could not happen because land was not as dense and heavy as sea floor, and could not sink into it. The only other explanation possible was that continents had once been joined, then had become separated and had drifted slowly apart for millions of years.

Little attention was paid to the idea of drifting continents until 1957-58 when most nations of the world cooperated in making all kinds of studies about the earth. The project was called the International Geophysical Year. So many new facts were learned that many began to believe that the early idea was right, and continents had moved. When the project was over, work continued.

Ten years later so much evidence had turned up that a special ship, the *Glomar Challenger,* was sent to sea to drill holes in ocean bottoms and bring up for study samples of mud and sediment millions of years old. By 1974 over thirty trips had been made. Each new discovery brought to light more proof that Gondwanaland did exist, and that it began breaking up about 200 million years ago.

End of an Ancient Continent

Gondwanaland began to break apart about the end of the Triassic Period. Great cracks appeared on both sides of Africa. As the splits gradually grew deeper, masses of land on each side of Africa separated from it. The large pieces became continents.

The fragments swung about, slowly changing position as they moved. India broke away completely and moved northward for a hundred million years. Finally it pushed against southern Asia with such force that the edge of the big continent rumpled up into the high Himalayan mountains. India is still moving slowly northward.

South America moved westward. As the split between it and Africa became wider, the Atlantic Ocean was formed. It stretched northward to where North America had begun to separate from Europe. The ocean was narrow here, for these two continents were still bound to each other by way of Greenland. This connection acted as a great passageway for millions of years, and animals of many kinds wandered from one continent to the other. Greenland did not become an island until long after the first mammals appeared on earth.

The Final Break

Australia and Antarctica remained linked together until near the end of the Eocene subperiod. Then Australia split away and began moving northward, carrying New

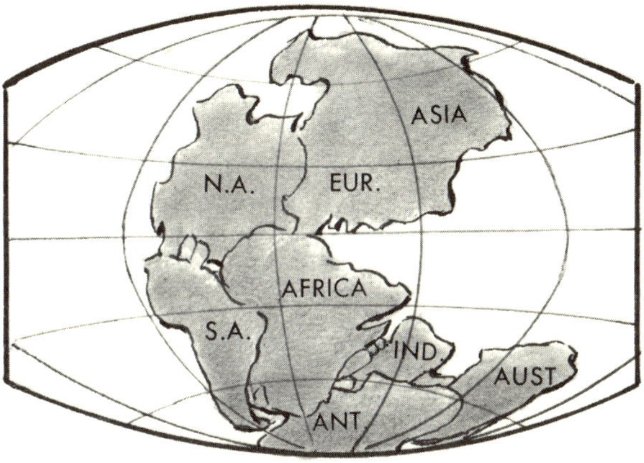

Pangaea, at top, cracked apart many millions of years ago and the pieces gradually drifted into the positions we know today. Australia, along with Tasmania and New Guinea, parted company from Antarctica and moved northward.

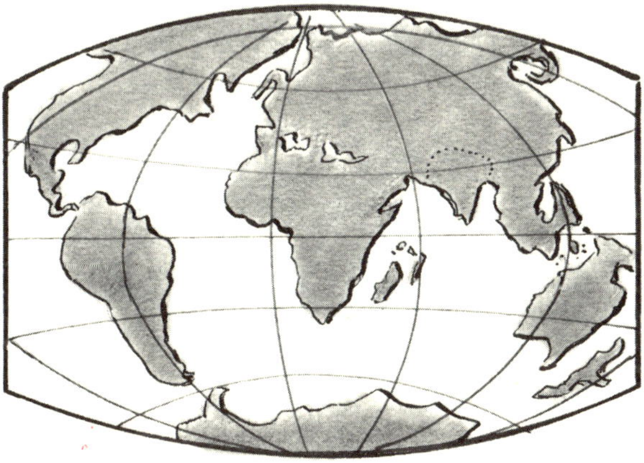

Guinea and Tasmania with it. About the middle of the Oligocene a chunk broke off one corner and became the island of Tasmania. New Guinea probably was under water until middle Miocene times. Both islands are really a part of Australia, for the land bridges that link them lie under shallow water.

At times these connections were laid bare and animals could cross back and forth to the mainland over dry land. This happened when glaciers covered a great part of the earth. Millions of tons of water were frozen and this caused the level of water in the sea to be much lower. Later, when the glaciers melted, water was set free and the sea level rose. Then islands were formed again. Such a change in sea level took place several times during the Ice Ages.

Today the sea level is "in between." If all the glaciers and ice at the two poles were to melt, the ocean level would rise so much that edges of continents and plains far inland everywhere would become covered with sea water. All of New York City would be on the ocean floor.

As Australia moved northward, Antarctica swung around until almost the center of it lay over the south pole. The long finger of its peninsula still curled up toward the southern tip of South America. Between the two stretched a curved ridge and chains of islands. We call this the Scotia Arc. Many of the islands are still there today.

Marsupials Go to Australia

For many years people have supposed that marsupials went to Australia from North America by passing through Asia. Both dinosaurs and mammals were able to go back and forth between the two northern continents. We know this is so because fossils found in both places are so much alike that the animals had to be either the same species or very closely related. This could not happen unless some kind of connection joined North America and Asia.

Studies show that the bridge over which the animals passed was at the Bering Strait. This is off the coast of Alaska where the tip of a peninsula juts out so far that the distance from our coastline over to Asia is about 80 kilometers (about 50 miles). Today the Bering Strait bridge is covered with water, but there is much evidence that a dry-land passageway existed between the two continents several times during the last 100 million years.

Since other animals passed over the bridge, there seemed to be no reason why didelphids could not have crossed also. Yet no fossils of marsupials have ever turned up in Asia, although bones and teeth of reptiles and placental mammals have been found in many regions of China and other countries. If any marsupials ever lived there, it seems strange that not a single fossil has ever been discovered.

There is no proof that early animals of any kind ever reached Australia by this route. Even though many rich fossil beds of dinosaurs and early mammals are known in Asia, not a single bone or tooth of these animals has been found in Australia. In fact, the oldest vertebrate fossils so far discovered there date back only about as far as the later part of the Oligocene.

Some of the early ideas about pathways to Australia are probably not true. Scientists who study oceans and islands have found no trace of any great land bridge that might once have connected Asia and Australia, then later sank beneath the sea. Animals sometimes "island-hop" from place to place by sailing on some kind of raft, such as a floating log, but it is not likely that any early creatures reached Australia by this method. Recent discoveries show that Australia was probably still attached to Antarctica during the Cretaceous Period and the early part of the Age of Mammals. This would place it thousands of miles away from its present position. No bridges or islands were near enough for walking over or "island-hopping."

A Southern Route

Many scientists today think that marsupials may have reached Australia from South America by way of Antarctica. This would be possible, because the tip of South America and part of Antarctica-Australia stayed close together long after the Atlantic Ocean opened up, and the other continents broke away from Africa. To see how this route could be used, get a globe (not a map) of the world and turn it so the south pole can be easily seen.

If the globe is colored so that water depths are shown, you can trace the Scotia Ridge, a long half-circle that stretches eastward from the tip of South America and then curves back again toward the northward-pointing finger of the Antarctic peninsula. There is much evidence that South America and Antarctica-Australia were sometimes separated and at other times united, probably by way of this bridge, for millions of years after the Age of Mammals, or Cenozoic Era, began.

Computer maps show that Australia's true southern edge can be fitted against the long coastline curve of Antarctica that we now know as Wilkes Land. This is almost opposite the peninsula. In the early part of the Age of

Mammals, Antarctica was still far enough from the south pole that it was not yet a bleak, frozen mass. Any animals that wandered into it could make the long trip overland and finally reach Australia, too. Such a journey might have taken any particular species some thousands or even millions of years.

Surprising Discoveries

Although scientists are more and more beginning to believe that marsupials followed this route, no one can be sure until many more fossils are discovered and studied. This is difficult in Antarctica because it is so bitterly cold in some places that the snow never melts and a few of the glaciers are up to five kilometers (three miles) thick. In other spots the bluffs are bare because strong, icy winds have blasted the snow away, and frozen particles of ice, hard as rocks, have eaten at the cliffs until crumbled piles lie at their bases.

Still, Antarctica is slowly giving up its secrets. Since 1959 scientists of several nations have banded together for study. Each nation has its own stations, but all share what they have learned with each other.

Some of the discoveries have been fantastic. Great seams of coal found on the frozen continent show that Antarctica was once near the equator, for coal is formed from ferns and forests in hot, steamy swamps. Vertebrate fossils have been discovered, too. The first one was part of an amphibian, a relative of salamanders and frogs, that lived about 200 million years ago. Peter Barrett, a New Zealand geologist, found it in 1968.

A year later, bones of a fresh-water reptile were found. This was a mammal-like creature with a heavy body, stocky legs, and a short tail. Fossils of this same reptile are common in South Africa and in India, and are further proof that these continents were once united.

No remains of mammals have yet been found in Antarctica, but the search for them has really just begun. It is quite likely that they, too, will be discovered before many more years have passed. If marsupial fossils are found in Antarctica, it would show clearly that the first land or freshwater animals in Australia could have entered by this route.

A Final Puzzle

The discovery of such fossils in Antarctica would also help us find the answer to another puzzle. The oldest marsupial fossils found in Australia so far date back only to the later part of the Oligocene epoch. Yet by this time half a dozen families had already developed. Some have since died out. Others, like the kangaroos and the meat-eating native cats, are very much a part of Australian life today.

Perhaps at least part of the development of these families took place in Antarctica, before the separation of the two continents during Eocene times. We know, from the great coal seams discovered, that the south-pole continent was once near the equator. When the climate was warm, large land animals could have lived there easily for millions of years.

Many of these ideas are still just guesses or theories. We will have to wait until much more of Antarctica is explored, and many more fossils are found in both Antarctica and Australia, before the true history of marsupials can be traced. At present, no one really knows where the marsupials of Australia had their beginnings.

5

Acrobats, Devils, and Quolls

Marsupials have lived in Australia for at least 30 million years, and probably for twice that long. During this time they have become adapted to all kinds of habitats. They live in mountains and deserts, forests and woodlands, in grasslands, rocky outcrops, and deserts. Since the white man came and began to change the countryside, some have learned to make their homes in towns and cities, but they still come out after dark to hunt for scuttling insects, tender leaves, or some other favorite food.

One large family of Australian marsupials is much like some of South America's pouched mammals in many ways. These are the dasyurids, or meat-eating marsupials. The word "dasyure" means "shaggy-tailed." All animals in this family have long tails covered with hair rather than with scales, but none uses the tail as a fifth hand.

Dasyurids seem to be more primitive than other kinds of Australian marsupials. This means they are less specialized, and that other marsupial families may have developed from some kind of dasyurid ancestor that lived millions of years ago. Their teeth and toes are much like those of didelphids, the oldest and most primitive of all marsupial groups.

Both dasyures and didelphids have several pairs of small biting teeth and long, sharp canines. The cheek teeth are those of animals that usually eat insects or other meat. Some dasyurids walk flat on their feet, and others use only their toes, but tracks made by all of them show that toes on the hind foot can be spread apart. This type of hind foot is not found in most Australian marsupials.

Dasyurids seem to be more intelligent than didelphids. They have alert, foxy-looking faces and the smaller ones have sharply pointed snouts with long, sensitive whiskers. The animals run rapidly, and can chase and catch their prey in a flash. Dasyurids will kill and eat any kind of animal that is not too large for them to tackle. Probably they eat some plant food at times; most carnivores do.

Small Acrobats

Many dasyurids are about the size of mice or small rats. These are often called marsupial mice. Probably the most common is the tuan or wamberger *(Phascogale tapoatafa).* It is found all over Australia but is rarely seen because it is quite shy, and is active only at night.

The shy tuan, called also a wamberger, is one of the marsupial mice.

The tuan has dark, smoky-gray fur, and is about the size and shape of a squirrel. Its tail hangs downward and the bottom half is covered with long, black hairs. Quite often the hairs are so fluffed out the tail looks like a bottle brush. This has earned for the tuan the nickname of brushy-haired marsupial mouse.

Tuans can be savage little fighters, but if they are caught and tamed early they make gentle pets that are easy to handle. They usually eat insects but sometimes feed on small mammals or birds. Farmers report that tuans raid chicken houses at times, but it is more likely that they are hunting for the rats or mice that often make chicken houses their homes. (True rats and mice are rodents, or gnawing mammals, and are placentals. We know from fossils that the first rodents reached Australia millions of years after the marsupials arrived. Many of them are now as much of a nuisance there as they are in other parts of the world.)

Tuans are tree-dwellers and are little acrobats. They can turn back-flips and land on a branch below as easily as they can climb up high again. Nests are usually made in hollow limbs and lined with grass or leaves. One tuan built an expensive nest. It made off with some paper money, then shredded it and used it for a lining.

Breeding takes place in winter, which in Australia is the same time as our summer. Three to six young are born at one time. The pouch on the mother is just a saucer-shaped area with a small rim around the edge. The young stay in the "pouch" for about four months, but are ready to have offspring of their own by the time they are a year old.

The Desert-Lover

The mulgara *(Dasycercus cristicauda)* is a dasyure that lives in hot, sandy, or stony desert regions. It is about the size of a small rat and has a thick-set body with reddish or sandy-brown fur. The nickname "crest-tailed marsupial mouse" is often used because a crest of stiff, black hairs stretches along the upper side of the tail from about halfway down to the end.

Each animal eats about one-fourth its weight in insects or other meat each day. Drinking water is scarce, but the mulgara is adapted to desert life. It obtains water from its food and loses very little of it. A mulgara's kidneys, like those of many desert creatures, produce a thick, concentrated urine.

These animals are often active during the day. One may dig a burrow, throwing pebbles and sand backward between its hind feet, or it may just lie basking in the hot sunlight. If some unusual object is placed nearby the animal is likely to nose and examine it curiously.

A hungry mulgara will attack prey savagely, killing it instantly with a bite at the nape of the neck. Then the animal may groom itself carefully before eating its meal. A mouse is devoured completely, from snout to tail tip, except for skin and bones. Yet a well-fed mulgara will allow a mouse to share its nest and not harm it.

The animals rarely fight each other. Instead, they "threaten." Each sits upright, facing the other, with mouth opened wide. Then each leans forward slowly. If their noses touch, both animals jerk back quickly. Then they slowly lean forward, "threatening" again.

"Hedgehog" Marsupials

Broad-footed marsupial mice, or antechinuses, are common over most of Australia. Some live in Tasmania, and others are found in New Guinea and nearby islands. They seem to prefer forests or woodlands, although many live in rocky country. All have ridged pads on palms and soles that aid them in climbing. Some species that hide in rocky crevices or caves can even run across a cliff ceiling for a short distance.

Antechinuses are mouse-sized and mouse-shaped, with short, tapering tails and long, pointed noses. The two most common species are the yellow-footed *(Antechinus flavipes)* and the brown *(Antechinus stuartii)*. Fur color in the yellow-footed one may range from pinkish-tan to coppery-brown, but the brown antechinus is dusky all over. The term "antechinus" means "like a hedgehog" and is used because the fur on all these animals has a bristly look.

Antechinuses breed during the Australian winter. Mating is a violent process and may last several hours. The male grabs the female's neck with his mouth and holds her tightly. Sometimes the wounds are so severe the female may die.

Most males die when the mating season is over, but females may live for another year. Three to twelve young are born at a time. They nurse about three months; six months later they are ready to produce families of their own.

Back from "Extinction"

Probably the most interesting antechinus is the dibbler *(Antechinus apicalis)*, which was first discovered in western

Australia in 1838. During the next fifty years a few more of the animals were seen; then they disappeared. Scientists searched for them, but could find no trace. Finally the animal was added to the list of extinct species.

Eighty-four years after the last one was seen, a man writing a book on wild flowers took a picture of a couple of small, brown, speckled marsupials he could not identify. They had short, hairy tails, small ears and sharp noses, and pale rings around the eyes. Dibblers had been found again, still very much alive.

Dr. Patricia Woolley, an Australian zoologist at La Trobe University in Melbourne, has now raised several of these small creatures that mated in the wilds and studied their habits and reproduction. No captive dibbler has produced young after mating while it was a captive, so a colony has not been set up.

Flat-Headed Pygmies

Four or more species of pygmy marsupial mice are known, but the smallest one is the Kimberley pygmy *(Planigale subtilissima)* which lives up in the Kimberley region of western Australia. From pointed snout to tail tip it measures less than ten centimeters (just under four inches) and over half of that is tail. This species lives in woodlands and grasslands, and wolfs down more than its own weight in food every day.

Pygmies are sometimes called flat-headed marsupial mice because their skulls are as flat as those of lizards. They can crawl into narrow rock crevices easily. During the dry season some even make their homes in cracks that are left in the mud when flood waters are gone. A new species, with brindled dark reddish-brown hair, has recently been reported.

The Galloping Wuhl-Wuhls

Wuhl-wuhls *(Antechinomys spenceri)* and some of their relatives have been known for years as jerboa marsupial mice. The animals have quite long hind legs and people thought they kangaroo-hopped, much as do the true jerboa mice of Asia and eastern Africa.

High-speed photographs made by Dr. W. D. L. Ride of the Western Australia Museum have shown this is not

The mouse-sized wuhl-wuhls have a good deal of curiosity and peer sharply at nearby objects.

true. Wuhl-wuhls move by galloping instead of by jumping. Using the hind legs as springs, the animals land on the forefeet, then quickly bring the hind legs around outside the forelegs and spring forward again. A racing wuhl-wuhl seems to be playing leap-frog with itself.

These animals are mouse-sized, with large ears and a brush on the end of the tail. They are full of curiosity and sometimes stand upright on their long hind legs to examine better some object nearby. Unlike many marsupial mice, wuhl-wuhls have a pouch, but it opens toward the rear.

Intelligence Tests

Dunnarts, or narrow-footed marsupial mice, are found throughout Australia and Tasmania, New Guinea, and islands nearby. Most are mouse-sized creatures with large ears and fine fur. About half the species live in dry regions and store food in the tail until it is quite fat. When food is scarce this store is used up, and the tail becomes thin again.

One of the best known of this group is the little fat-tailed dunnart *(Sminthopsis crassicaudata)*, which can be found over most of Australia. It is easily recognized by its short tail and dark patches on the ears and head. Marsupials have long had the reputation of being more stupid than placental mammals, but this small creature is helping to show that this idea is not necessarily true.

Robert J. Kirkby and Alan C. Preston, of the Psychiatric Research Unit in Rozelle, New South Wales, have compared the behavior of the dunnart with that of two groups of ordinary placental mice, one white and one colored. Each animal was placed by itself in a T-shaped box with the bottom marked off in squares. Some odd object,

The fat-tailed dunnart, found almost throughout Australia

such as a red plastic cup or a bulldog paper clip, was placed in one or both arms of the T.

The animals were left in the boxes ten minutes. During this time tests were made to check such facts as how many squares the animal crossed, how much it jumped around exploring, and how soon it got bored and did nothing. Each animal was tested several times.

All the black and white mice acted very much alike, but the dunnarts made better scores on every test. They covered more squares, which showed they were alert and active. They jumped around more than ordinary mice, finding and examining the strange objects. Also, they got bored more quickly. When they had explored the box thoroughly, they settled down and groomed themselves.

Studies made with dunnarts are only a beginning. Tests with other marsupials may show that their ability to learn is much like that of placental animals. Whether hogs, horses, or human beings are being tested, some are smarter than others. We can no longer take it for granted that all marsupials rate below placentals in intelligence.

The Quoll and Its Relatives

Larger species of dasyurids are usually known as "native cats." Most members of this group are cat-sized or a little larger, but they are not at all related to true cats. They are lean, long-legged, and active. The fur may range from olive-brown to quite dark in color, and is covered with white splotches. One dasyure, the tiger cat, has spots to the end of its tail.

Probably best known is the quoll, or eastern dasyure *(Dasyurus viverrinus)*. It is found on both the Australian mainland and Tasmania. The animals sleep by day, curled into a ball with ears folded down, and creep out at night to look for food. They eat lizards, rabbits, fish, or whatever is at hand. The animal stalks its prey carefully. When close enough, it leaps and kills with a bite at the back of the neck. It eats quickly, and is soon off again, searching for more food.

The quoll is one of the few Australian marsupials that, like the American opossum, produces more young than it can take care of at one time. Ten to 18 may be born but some must die, for only six nipples are in the pouch. Weaning begins after eleven weeks, and the young quolls can catch

Some of the dasyurids are called "native cats" by the Australians, though they are not related to cats.

their own prey and care for themselves by the time they are four and one-half months old.

Affectionate Devils

Another flesh-eater, well-known but often feared, is the stocky, short-legged Tasmanian devil *(Sarcophilus harrisii)*. It has shaggy, black hair, a large head, and powerful jaws and teeth, and looks much like a small, clumsy bear. Sheepherders say it threatens their flocks, but research has shown that only weak or sick sheep become its prey.

These animals were first discovered in 1808 by Robert Harris, a Deputy-Surveyor, who kept two of the animals chained in a barrel for several months. He reported they were savage, vicious creatures and "uncontrollably wild." Other captive devils, if given running room and plenty to eat, seem quite content. Fights are rare although a cornered

In spite of its name, the Tasmanian devil can become an affectionate pet.

animal may growl and threaten with a sort of snarling cough.

The late Mrs. Mary Roberts of Beaumaris, Tasmania, kept and bred Tasmanian devils as pets. She found them to be mischievous but affectionate and very clean. Each groomed its face and fur daily, licking its paws and cleaning its coat.

Fossils of these animals have been found in Australia, but none has lived on the mainland for several hundred years. Remains dated six or seven hundred years ago have been discovered in Victoria. Devils are found mainly in wild, rocky, forested areas with dense thickets and underbrush.

Dens are made under stones or in old tree stumps. The pouch, with four teats in it, is turned toward the rear, and the opening can be tightly closed as the mother makes her way through thick brush. Young ones are weaned when six months old but do not breed until about the second year. Tasmanian devils may live for six or seven years.

The Ghost Wolf of the Rain Forest

The most fascinating of all Australia's carnivorous marsupials is the pouched wolf *(Thylacinus cynocephalus)*. Its scientific name means "pouched dog with a wolf head." The animal is so much like Borhyaena, the South American meat-eater that lived during the Miocene subperiod, that some scientists think perhaps both animals should be placed in the same family.

Others disagree. They think the resemblance is more likely to be a case of parallel evolution. This means that the two animals grew to be much alike because they lived under the same sort of conditions and reacted to them in much the

The sturdily built pouched wolf is a fast and cunning predator that seems quite fearless.

same way. Both are large-sized, and both are meat-eaters that chased and killed their prey. Their teeth and bones are much alike, too, which shows that they had similar habits. Probably both came from the same common stock, a small, opossum-like didelphid that lived over 100 million years ago.

The pouched wolf goes by several names. Sometimes it is called the Tasmanian tiger because its dense fur has chocolate-colored stripes across the back and hips. The rest of the animal is a yellowish or reddish brown. However, the often used name "thylacine" may turn out to be best, because the animal is really neither a wolf nor a tiger. It is like other dasyurids in many ways, but is so different that it is usually placed in a family by itself.

Thylacines are not as speedy as dogs, but they run on tiptoes as dogs do. The animals are faster and more cunning than other marsupial carnivores, and seem to be fearless. They have never been known to harm humans, but a cornered one will turn and face a pack of dogs, growling hoarsely and ready to fight.

The natural food of thylacines is wallabies, or small kangaroos, and other small mammals, birds, and lizards. The animals may hunt singly or in pairs. They trot after their quarry until it grows too tired to run farther, then they move in, pounce on the victim and bite through the skull, killing it instantly. Bloody parts such as the liver, which is loaded with vitamins, are devoured first. A Tasmanian devil may eat animals dead for several days, but a thylacine feeds only on fresh meat and never returns to its kill.

Too Many Bounties

During the last century thylacines were common in Tasmania, although none has been found on the Australian mainland for well over 2000 years, or in New Guinea for 10,000. When white settlers first moved into Tasmania and began raising sheep, farmers reported that thylacines raided their herds. By 1840 a Land Company began paying a bounty for each scalp.

A few years later the government took over. Between 1888 and 1909 over 2000 scalps were turned in, and $25 paid for each one. One man claimed to have killed 24 of the animals in a single day.

The number of bounties paid reached a peak in 1900, but by 1910 payments had dropped to zero. During the next ten or twenty years only a few of the animals were found, and they were sold to zoos. The last one of these died in the Hobart Zoo in 1933.

Four years later the Tasmanian government put the thylacine on the list of wholly protected species. Anyone who killed one of these creatures now became subject to a $500 fine, but the action came too late. There is no proof that any thylacine has been seen since that time.

Reports come in every few years that one has been sighted, but no one has yet been able to prove it. Since 1938 several teams of scientists have joined in a search to see if they could possibly find and save some of the creatures. They wanted to place them in special reserves where they could be protected and studied. No search has been successful.

The last real trace of one of these animals was reported in 1966. A farmer on the west coast of Tasmania found a lair with a few hairs in it that were identified as belonging to a thylacine. The location of the lair was kept a secret, and snares set, but the animal was too wary to be caught.

Perhaps thylacines have gone the way of the American passenger pigeon, and a living one will never be seen again. Or perhaps a few are still alive in some part of the craggy, mountainous regions of Tasmania that are still wet, wild, rugged, and almost unexplored. If any thylacines do exist they seem to have learned, as many other animals have, not to trust man.

6

Five Odd Families

Bandicoots are peculiar creatures. Their teeth are similar to those of didelphids and dasyures, and show that meat makes up a good part of their diet. Young bandicoots have the sharp, needle-like teeth of insect-eaters but in older animals the teeth become worn and dull.

Yet legs and feet in these animals are not at all like those of dasyures. Instead, they resemble those of kangaroos and most other Australian marsupials. The hind legs are large, the feet flat, and the main toe is the fourth one. Beside it is a "twin toe." This is really the second and third toes but they are so closely joined they look like one toe with two claws. Such toes are used for grooming. They act as a comb when drawn through the fur.

Perhaps the oddest thing about bandicoots is that a simple kind of placenta develops as the embryo grows. Eggs of marsupials have a heavy membrane around them that is much like the white lining in the shell of a hen's egg. This covering keeps the embryo from becoming attached to the uterus wall, but liquid food in the uterus can reach the embryo by seeping through the membrane.

In most marsupials, the membrane breaks after a while, and the embryo becomes loosely fastened to the uterus wall. During this time it receives food from a yolk sac that grows below its body. Within a few days the yolk is used up. Then the infant marsupial is born.

The membrane in bandicoots breaks soon after the

embryo starts to grow. Then a true placenta is formed. It is very simple and does not last long. When the infant is born it is still just a minute embryo, no larger than any other newborn marsupial.

"Nose" Animals

Because they have such long, pointed snouts, bandicoots are sometimes called "nose marsupials," although the scientific family name is peramelids. This name means "badger-like" and refers to the habit all these creatures have of using their sharp claws to dig for food. They seem to be quite fond of such animals as earthworms and insect larvas, but bulbs and plant roots also form a part of their diet.

The common long-nosed bandicoot *(Perameles nasuta)* is found in eastern Australia, where it lives in forests and woodlands. It is about the size of a cottontail rabbit and has harsh fur that is gray to brownish-gray in color. The ears are long and pointed, and the tail hairy.

Many long-nosed bandicoots live near town and pester gardeners by rooting for insects. If disturbed, they take off at a gallop. Sometimes one will jump straight up in the air, then turn and race off in another direction.

Pouches in bandicoots open to the rear, between the mother's hind legs. There are eight teats, but only two or three young are born at one time. As in other marsupials, nipples become larger and longer as the infants grow, and do not shrink back to normal size until some days after the young have left the pouch. New infant bandicoots usually appear before this has happened, so each must hunt for one of the other nipples that is still small enough to fit its mouth.

A closely related species, the spiny bandicoot *(Echym-*

ipera rufescens) is common in New Guinea but occurs in Australia only in the tropical rain forests at the northern tip of Cape York. Probably some of the animals crossed from Australia to New Guinea long ago at a time when Torres Straits formed a land connection to the mainland. Later, some migrated back again.

The "Tailless" Bandicoot

Probably the oddest member of this family is the pig-footed bandicoot *(Chaeropus ecaudatus)*. It is not quite as large as a rabbit, and its forefeet look like those of a pig because only the second and third toes are developed. The hind foot is mostly a large fourth toe. These animals have slender, stiltlike legs and are much better at running than at digging.

The first specimen, caught in 1836, lost its long tail when pulled from a tree by its captors. Zoologists gave the animal a scientific name that meant "the pig-footed bandicoot without a tail." It was years before the mistake was discovered, because naturalists who went to hunt more specimens used a picture to show their hunters what kind of animal they wanted. The hunters captured more bandicoots easily, but carefully cut off all their tails before turning them in, in order to be sure the animals looked just like the picture.

Little is known about these small, deerlike animals, for none has been definitely seen since 1907. Some were reported in 1926, but there was no proof these were really pig-footed bandicoots. Like many other marsupials, this one also seems to be extinct, or nearly so.

The bilby's large ears, aside from helping it to hear well, probably serve also to keep it cooler by radiating heat efficiently.

The Bilby

Rabbit-eared bandicoots *(Macrotis lagotis)* are graceful creatures with long ears and soft, silky fur. Most are cat-sized or a little smaller. The tail is black and white, with a white crest, and it is often carried upward as if it were a banner. Bilby is a common name for it.

Bilbies are the only bandicoots that dig deep, winding burrows. They live in dry, inland areas and hide in the burrows during the heat of the day. They probably make use of their big ears for cooling off, too, as our western jackrabbits do, by radiating heat from the large, thin surfaces.

Bilbies eat both meat and insects. When asleep they squat on their tails with heads tucked down between the forepaws, and the long ears folded over their faces. The animals were once common, but as more land has been opened for agriculture they have retreated to dry, out-of-the way places. None has been found near populated areas for 50 or 60 years.

The Termite-Eating Numbat

A peculiar little Australian marsupial, commonly known as the numbat *(Myrmecobius fasciatus)*, is a rat-sized creature with a stout body and thin legs; all its toes are separated and have claws. It feeds on insects. For these reasons some zoologists think the numbat should be placed

A numbat eats thousands of termites every day.

with the dasyurids as a special subfamily. Others disagree because the animal is different from dasyurids in so many other ways they believe it should be in a family by itself.

The numbat has a long, pointed muzzle and unusual markings. Its fur is coarse but silky, and is a reddish brown color with white hairs sprinkled through it. Across the back and hips are white bars with dark streaks in between. A dark cheek stripe, with white above and below it, runs through each eye from the ear to the nose.

The scientific name of the numbat means "banded anteater," but the favorite food of the animal is termites. Ten to twenty thousand a day are eaten. The numbat has a round, sticky tongue almost half as long as its body. When the animal is searching for termites in chunks of rotten wood the tongue is flicked rapidly in and out the tunnels and galleries. Small termites are swallowed whole. Larger ones are chewed with some of the creature's 52 teeth.

Numbats are found now only in forests of southwest Australia where most of the trees are a kind of eucalyptus known as wandoo. Termites attack both living trees and fallen branches. The numbat feeds during daylight hours and sleeps at night curled in a hollow log. It crawls in head-first, tucks its tail under its body, then fills up the opening with its broad rump. No enemy can dislodge it.

The numbat is harmless. It has never been known to bite or to fight. Instead, it is very valuable as a termite-eater. The species is not now on the endangered list, but as human settlements spread the wandoo forests are likely to disappear. The wood is good for charcoal and the leaves yield an oil used in many industries. Perhaps, when wandoo forests themselves are in danger, parks or reserves will be set up where these animals can be protected.

One of a Kind

The marsupial mole *(Notoryctes typhlops)* is so different from all other marsupials that it has no known relatives. It might be distantly related to dasyurids, or perhaps to bandicoots, but so little is known about it that no one is sure. The animal looks and acts more like the golden mole of Africa than it does any creature in Australia. Yet it is a marsupial and the African mole is a placental. The likeness between the two is another remarkable case of parallel evolution.

The mole is small enough for you to hold in your hand. It has fine, velvety fur that may vary from a deep cream color to a rich gold. Small, blind eyes are hidden under skin and muscle. No outside ear flaps show, but two small holes, which can be closed, lead to inner ears. The nose has a horny shield over it, and the short, naked tail looks like a leathery cone covered with rings.

Marsupial moles live in desert country. The forelegs are short and have great, flattened claws on the third and fourth toes that can be used as scoops. The animals spend most of their lives tunneling through sand, flipping it backward with their hind feet. The burrows cave in behind them.

Moles eat earthworms and other creatures that live in sand and soil. After digging a while a mole may come to the surface for a short time, but life above ground is not easy. The animal shuffles along, flat on its belly and pushing its body with its hind feet. Soon it begins digging again.

Marsupial moles have been kept as captives for a while several times so their habits could be studied. Such confined

The marsupial mole is very different from all the other marsupials and spends its life tunneling through the sand.

animals seem to be highly excited and always hungry. They drag themselves forward in a headlong rush, gobble any earthworms or grubs they can find, then struggle onward again. At times a mole may fall asleep suddenly, without any warning. Soon afterward it awakens just as quickly and tears off again.

Marsupial moles carry one young in the pouch at a time. Nothing else is known of their breeding habits.

A Twin-Toed Plant-Eater

Australian marsupials can be divided into three great groups: meat-eating dasyurids and their relatives; bandicoots and other "in-betweens"; and animals that feed on plants. This last group is by far the largest, and includes several families.

The wombat family belongs in this last group. These animals are stocky, heavy creatures with big heads, short legs, and practically no tail. A fully grown wombat looks much like a cross between an overgrown woodchuck and a small bear.

In many ways the common wombat looks like a miniature bear, as its species name, Vombatus ursinus, *suggests.*

The common wombat *(Vombatus ursinus)* lives in rocky or mountainous forested regions in southeastern Australia and Tasmania. It has a shiny black nose, small ears, and short, coarse fur that may be either a grizzled gray or brown. Other species, known as hairy-nosed wombats, live on plains.

Wombats have teeth much like those of beavers. Several pairs of cheek teeth are present, but no canines. Only one pair of front teeth is in each jaw but these are large and strong, and are softer on the back side. The softer part wears off more quickly and this keeps the biting edges chisel-sharp. None of a wombat's teeth has any roots, and all keep growing as they are worn away.

Toes of wombats are armed with strong claws and the animals are powerful diggers. They make branching tunnels that connect with each other and sometimes form a great network. A digging wombat sits on its rear end and hacks out the earth with its forepaws, pushing it to one side. Then the animal backs out of the tunnel, kicking out dirt as it goes.

Nesting chambers are lined with grass and are far underground. The pouch points backward and contains two teats, but only one young is born at a time. It stays in the pouch over nine months, then waits another year before it has infants of its own. The rear end of a mother wombat with the big head of a young one looking out from between her hind legs is a rather startling sight.

Grasses, bark, and roots are the favorite foods of wombats. Farmers have no use for the animals because they sometimes tear large holes in fences and feed on the crops. In addition, the long burrows make good hideaways for the thousands of rabbits that plague the countryside.

Wombats are inquisitive and friendly, and make good captives. If kept as a pet, one may follow its owner around, play with children and, if allowed to do so, curl up on a handy lap and go to sleep. If it is hungry, a wombat begs for food with queer little noises. Wombats are long-lived. One in a London zoo was 26 years old when it died.

Australia's Sweetheart

The cuddly koala *(Phascolarctos cinereus)*, sometimes called the Australian teddy bear, is one of Australia's most-loved animals. Many of the aborigines (people who lived in Australia for thousands of years before white men came) thought that spirits of lost children lived again in these animals. Perhaps the childlike cry of a hurt koala helped start this belief.

Koalas spend all their lives high in eucalyptus, or gum trees. They look much like small, chunky bears with fuzzy, tufted ears, bright black eyes, and a shiny black nose. The thick fur is ashy gray above with a tinge of brown, and whitish on the underside. The tail is a short, rounded stump.

All four feet of koalas have strong claws that are useful for grabbing and holding onto trees and branches. In the hind feet the first toe is opposable, just as our thumbs are. The second and third hind toes are "twins," joined by a common skin. In the front paws the first two toes are opposable to the other three.

Gum Tree Specialists

Koalas smell like cough drops because they never eat any other kind of food than leaves of a few species of eucalyptus. Oil from these trees is commonly used in making cough drops and certain other medicines. When tender shoots on one tree are gone, the animal comes down and clumsily crosses the ground to another tree. It can climb smooth trunks, scrambling upward in short jumps and holding on with its claws.

At certain times of the year, leaves of some gum trees produce a strong poison. The koala must then find another kind of gum tree to live in for awhile, or die. Unless tests are made in a laboratory, it is impossible to tell when the leaves become poisonous. Only a koala knows when to make the change.

The pouch of these animals opens toward the rear and has two nipples deep inside. A young koala stays in the pouch for eight or nine months, then rides on the mother's back, clinging to her fur, for several months more. During this time the mother may turn a misbehaving cub over her knees and spank it soundly. Sometimes she also hugs the baby to her while she rests, much as a human mother might do.

For years scientists wondered why a tree-climbing mother, that even sat in a tree crotch to sleep, had a pouch opening that was upside down. Then they found that the mother weans her child in a most unusual way. First she empties the droppings, or waste material, from her intestines, then she releases a thin, greenish "soup" made of partly digested gum tree leaves. The infant, snug in its pouch, licks the soup from its mother's anus. In this way its body gradually learns to take care of the kind of food that the adult koala will eat all its life.

A young koala rides on its mother's back for several months before walking about by itself.

The Koala Family

Koalas are quite different from most other marsupials. For many years they were thought to be most closely related to wombats, and were sometimes included in that family. Some years ago, because their teeth are much like those of the ringtail phalangers, they were placed with this group. Scientists now question both these relationships.

Dr. J. A. W. Kirsch, at the University of Western Australia, has studied the blood serum of 80 different species of Australian marsupials, and three species from South America. The tests he used were something like those made in hospitals to see whether or not two kinds of bloods match. If the two specimens of blood are alike, a transfusion can be given. If they do not match, a transfusion could cause death.

Dr. Kirsch found that koalas were, after all, more closely related to wombats than to ringtailed phalangers. Yet they are not quite like wombats either. Perhaps the best solution to this problem is to put the animals in a separate koala family, by themselves.

Senseless Slaughter

About one hundred years ago millions of koalas lived in Australia. They are such gentle, trusting creatures that usually they will neither run nor fight. Because they made easy targets, thousands were killed for "sport." Often several shots had to be fired before an animal would fall from the tree, because a hurt koala will wail loudly and hang on tightly.

Later it was found that the beautiful, soft fur of koalas would bring in good money. Wholesale killing then began in earnest. In the year 1924 over two million koala pelts were sold overseas.

Soon no koalas were left in the states of southern and western Australia. In Queensland, where the animals still lived, an open season was declared in 1927. This meant that anyone could kill all the koalas he wanted to, all year long.

The public was horrified. The outcry was so great that one state after another changed its laws. Now, in all parts of Australia, including the state of Tasmania, no koala may be killed. No one is even allowed to keep a koala as a pet without special permission from the government. The animals are special wards of the state.

The Fisheries and Wildlife Department now breeds koalas in reserves, then sets them free in places where their special eucalyptus trees grow, and they can live. Sometimes the wardens stop off at schools along the way to show the children how harmless and pretty the animals really are. The koala is one of the few animals that was almost wiped out, but now has a second chance to live.

What will happen to the others?

7

Phalangers and Possums

Australia was not really discovered until nearly two centuries after Columbus made his first trip to the new world. Several captains scouting the South Seas had seen the coastline here and there, but no one had realized that behind these coasts lay a great southern continent.

Captain John Cook, who led the first expedition to Australia, reported it would be a good place for new settlements. He told of large forests, sheltered bays, and odd animals. To his amazement, some of the creatures he saw hopped on their hind legs, but went faster than a greyhound. One animal he captured reminded the men so much of the American opossum that they dubbed it the Australian possum.

Later, when settlers moved to Australia, they started calling most of these strange pouched creatures by familiar English names. They knew that native cats, marsupial mice, and pouched wolves were not true cats, mice, or wolves, but this was an easy way to name and remember the new animals.

The name "possum" was adopted, too, following John Cook's lead. These furry little tree-dwellers of Australia seemed to be much like the American opossum, especially since animals in both groups carried their young in a pouch. Early settlers were not biologists. It was easy for them to believe that all opossums probably were very closely related, whether they lived in America or Australia.

Two Families

Many intelligent people today who know little about animals make the same mistake. They take it for granted that there is an opossum family which includes creatures on both sides of the Pacific Ocean. As a matter of fact, these two groups of animals, American and Australian, are no more closely related than are skunks and bears, or pigs and camels. Only the American opossums belong to the didelphid family.

Australian possums are phalangerids. Members of this family are different from didelphids in many ways. They eat plant food, for example, instead of worms or insects or whatever is at hand, and their teeth are fitted for this diet. The lower canines are missing; there are fewer biting teeth, and the first ones are long and stout.

The feet of phalangerids are much like those of the koala. There are strong claws on all toes of the front feet, and the first two toes are opposable to the other three. On the hind feet the first toe has no claw and the second and third are joined as "twin toes." The fourth and fifth hind toes are largest.

Ringtails

The animal that Captain Cook captured was the first Australian marsupial to be identified. We know it now as the common ringtail possum *(Pseudocheirus peregrinus)* or the gray Queensland ringtail. It is smaller than a cat and has soft, woolly fur. The tail, which is naked on the lower

The first marsupial to be identified in Australia was the ringtail possum.

end, is kept curled into one or two rings. When it climbs, the animal twists its tail close around a branch until its paws have a good hold on another limb.

Ringtails are nest-builders. They make a hollow ball of twigs and leaves and often line it with fern fronds. Sometimes nests are hidden in dense scrub near the ground. Others are built so high in trees that they sway when the wind blows. Ringtails feed at night and seem to be especially fond of the leaves and blossoms of peppermint gum trees.

One of the prettiest of the ringtails is the green possum *(Pseudocheirus archeri)* which is found in northeastern rain forests. Its fur is golden green, with two stripes running down its back. Dr. Hobart Van Deusen, of the American Museum of Natural History, examined hairs from one of these animals under a microscope and found they were a solid, clear yellow color with black spots. When white sunlight shines on them the mixture gives a greenish-yellow color.

The green mountain ringtail of New Guinea is such a close relative of this species that the Australian green possum is thought to have migrated back into Australia from the northern island long ago.

Striped Possums

Another possum that probably developed for some time in New Guinea, then later returned to the Cape York peninsula of Australia, is the striped possum *(Dactylopsila trivirgata)*. This animal has black and white stripes that run from its nose to the tip of its tail. It can be almost as smelly as the striped skunk of North America, though it is not able to spray its scent. Instead, it growls and shrieks at its enemies.

The striped possum is much like the small aye-aye of Madagascar, off the eastern coast of Africa, but the two animals are not at all related. This is just another case of parallel evolution. Both creatures have sharp, chisel-shaped biting teeth that are used to tear open the bark of trees. They tap their finger-like front toes on tree bark, then listen carefully for any noise made by some adult insect or larva moving about underneath. The possum then thrusts into the crack a long, thin fourth finger with a hooked nail on the end, and digs out any termites or larvas hiding there. The aye-aye does the same but with its third finger.

Gliders

"Flying" phalangers have developed in three different branches of the Australian possum family. All three look much alike, but they are not closely related. The scientific name of the largest one, the greater glider *(Schoinobates volans)* means "flying rope dancer." Teeth, skull bones, and blood tests show it is more closely related to ringtails than to other gliders.

Greater gliders live in the wet eucalyptus forests of eastern Australia. Their fur is soft and silky, and so long and dense that the animals sometimes look shaggy. Colors range from black through smoky or brownish gray to creamy white.

Gliders do not really fly, but jump from high places and sail through the air much as our flying squirrels do. A skin membrane covered with hair stretches from the forearms to the hind legs. When the animal jumps, it spreads its legs wide, and its flattened body is held up by an air cushion underneath. The long, furry tail acts as a balancer.

A greater glider can sail up to 100 meters (well over 300 feet) from a high tree to the foot of another one. It climbs the second tree quickly in a sort of gallop, then leaps

Greater gliders get about much as our "flying" squirrels do.

into the air again. A young tree glider just out of the pouch rides on its mother's back as she soars through space. The animals are graceful in the air but quite clumsy when they waddle across the ground.

Spitfire Sugar Gliders

A smaller "flying" possum *(Petaurus breviceps)* is known as the sugar glider because of its fondness for sweets. Its food is usually leaves or flowers, occasionally insects.

These animals have soft, silky gray fur, with dark markings on the face and a dark stripe that runs from the nose to the rump.

Sugar gliders are found in Tasmania and New Guinea, as well as eastern Australia. They make nests high in trees, and the animal often hangs head downward, gripping a branch with its hind feet, while it gathers material. Twigs and leaves are bitten off, passed rapidly with the paws to the tail, then carried in the curled tail tip as the animal races along high tree limbs.

Usually a dozen or more animals share a nest. This includes a male, two or three females, and sometimes several sets of offspring. Members of the community mark each other by rubbing secretions from scent glands onto the fur of the others. They also signal each other with warning calls, grunts, and quiet hisses.

Gliders in a community rarely fight one another, but if an animal with a foreign smell wanders too close it may be pounced on, and scratched and bitten before it can escape. In spite of their tempers, the creatures make good pets. One may sleep contentedly all day in its master's pocket, then spend the night hours hunting for food.

The Fairy Possum

A close relative of the sugar glider is a curious little squirrel-sized phalanger known as Leadbeater's possum or the fairy possum *(Gymnobelideus leadbeateri)*. It was discovered a century ago; then it disappeared from sight. Except for some stuffed museum specimens, no one knew how the creature looked.

In 1961 a survey was made of districts near the National

Museum of Victoria. One of the photographs taken showed three small, furry animals, each with a striped face and back, and a long-haired, bushy tail. After fifty years the lost possum had been found again.

Several of the animals have now been captured for study, so scientists can learn more about their habits. They live in cool mountain forests that are dense and wet. Probably they eat insects, and breed in late spring but, as yet, no one knows for sure.

Australia's Most Widespread Phalanger

The common brushtail possum *(Trichosurus vulpecula)* is found all over Australia. Some live in woodlands, others in half-desert regions or rocky dens, and many make their homes in attics of houses in suburbs. Like the American opossum, the common brushtail can live anywhere and eat anything. Although as large as small foxes, the animals are not very well known, for they roam by night and sleep by day.

Brushtail possums are sturdy creatures with thick, woolly fur that ranges from black to gray. A pink, finger-like tip is on the end of a long, bushy tail. These animals can use the tail as a fifth hand. Scent glands are on the chin and chest, and the animals make "scent posts" by rubbing the secretions on stumps and other objects.

No one knows what "message" is left by this means, for the adults live alone and do not fight over territory. If two adult males meet, they stop a moment and peer at each other with ears erect. Then each goes its own way.

A brushtail will defend its den if disturbed. It first makes clicking noises, then opens its mouth and hisses. If the intruder does not leave, the possum is likely to rise on its hind legs and screech, then fight viciously.

The common brushtail is easily tamed, and makes an affectionate pet. One will soon learn to answer to its name and come when called. Owners of hardwood forests like to have the animals around because one of the possum's favorite foods is leaves of the mistletoe. This is a parasitic plant that damages trees. With the mistletoe gone, the trees are in much better condition.

The common brushtail possum can be tamed as a pet and can answer to its name.

A Rare Relative

One of the rarest of the western phalangers is the small scaly-tailed possum *(Wyulda squamicaudata)*, a close relative of the common brushtail. The few that have been found live in rocky, forested country. Brownish gray, thick fur covers the body and base of the tail. The rest of the long tail is naked and scaly. The animals are quite gentle and thrive on blossoms, fruit, and nuts.

Invaders from the North

Cuscuses, also close relatives of the common brushtails, are round-headed, odd-looking marsupials that are found on more Pacific islands than any other pouched mammal. They are common from Celebes in the west to the Solomon Islands in the east. The few species on the mainland of Australia are thought to have returned again from New Guinea during the Ice Ages, several thousand years ago. Most of the cuscuses in Australia still live in the Cape York peninsula.

Probably best known is the spotted cuscus *(Phalanger maculatus)*, a chunky creature not much larger than a good-sized cat. It has a yellow nose, ears so small they seem almost to be buried in the fur, and large, protruding eyes with yellow or orange rings about them. The eyes themselves may be red, orange, yellow, brown, or even bluish green.

Its fur is dense and woolly, and males especially may be any color from a creamy white through yellowish, rusty, or grayish green to almost black. White spots are usually scattered over its coat. The less colorful females are grayish

Cuscuses spend all their time in trees.

or honey-blond with darker spots. Colors change with age, with the health of the animal, and with the place where it lives. Several color combinations are common in the same litter.

Cuscuses live in dense forests and spend all their time in trees. During daylight hours they sleep sitting in tree forks, held in place by their long, tightly-curled tails. A patch on the rump of each animal grows lighter with age because the fur is worn down with the long hours of sitting.

Feeding begins after darkness falls. Cuscuses eat plants, insects, bird eggs, and even small birds. An animal ready to feed sits on a branch, attaches itself securely with its tail and hind feet, then holds the food with its forepaws while it eats.

Cuscuses are full of curiosity and make interesting (but smelly) pets. One scientist who kept a cuscus in his room reported that the animal snarled, barked, and clawed at first but soon became tame. Then it investigated everything in sight. Another scientist who kept cuscuses for a while found that his suits still smelled musky after they had been sent to the cleaners twice.

Pygmy Possums

Pygmy possums are about the size of mice but have tails that curl at the tip like those of ringtails. Pygmies live in woody country and are active at night or on cloudy days. The common eastern pygmy possum *(Cercartetus nanus)* is grayish or fawn-colored above and slaty below. Its fur is soft and dense, but its ears are naked.

Pygmies eat some plants, but most of them prefer such food as insects and lizards. The animals sit to eat, holding the food with the forepaws. As insects become scarce in the fall, fat is stored in the tails. During cold weather the pygmies spend much of the time sleeping in tree holes or abandoned bird nests.

Several pygmies usually sleep together, curled into a ball. The nose is held against the chest, the tail is in a tight curl, and the thin ears are folded down. Real hibernation, a deep sleep that lasts most of the winter, does not occur.

The pygmy sleeps for a week or so, then awakens and hunts for food again. During the time it is asleep its body temperature is about equal to that of the air around it.

A Fossil Surprise

One of the most interesting relatives of the pygmy possums is the Burramys possum *(Burramys parvus)*. This animal is so new to science that it has no common name. Fossil bones of these creatures have been known for nearly a century, but no one suspected any such animals were still alive until August, 1966, when two scientists went skiing in the Australian Alps of eastern Victoria. Upon returning to their ski hut Dr. K. Shortman, a biochemist, and D. F. Jamieson, a chemical engineer, found a small possum sitting in the kitchen. It looked the men over curiously, but showed no sign of fear.

The two skiers fed their new friend cheese and apples, then took him to the laboratories of the Fisheries and Wildlife Department. Scientists there were amazed to find that this was a male Burramys possum, an animal thought to be extinct for over 20,000 years. The scientists cared for the possum, and immediately started a search for a mate. None was found, and the lone little animal died nine months later.

Several populations of Burramys possums have now been found in both Victoria and New South Wales. Captive colonies have been set up at the Division of Wildlife Research in Canberra, the capital of Australia, and in the Taronga Zoo in Sydney, New South Wales. With studies of breeding habits and other kinds of behavior, we should soon know more about this small animal that has "come back from the past."

The pygmy phalanger known as the flying mouse is so small that six of them can be held in one person's hands.

The Pygmy Glider

One of the pygmy phalangers is sometimes known as the flying mouse *(Acrobates pygmaeus)* or pygmy acrobat. It is so small that a half dozen could be cupped in your hands. The figure of this little glider appears on the Australian one cent piece.

Pygmy gliders have soft, thick fur that is grayish brown above and white on the underside. Long hairs fringe the skin membranes between the legs, and a flat fringe of hairs on the tail give it a paddle-like appearance. Because of this the animals are sometimes known as feathertail gliders.

Pygmy glider nests are made of gum leaves, high in some hollow branch. The animals sleep here during the day and at night hunt for soft-bodied insects such as ants or termites, or lick the sweet nectar from blossoms of eucalyptus or other native trees. When the weather is cold the animals go into a state of half-hibernation and sleep for several days.

Honey Suckers

Called by Australians the noolbenger, the honey possum *(Tarsipes spencerae)* is another marsupial that is often put in a family by itself because it seems to have no close relatives. The animal is not much larger than a pygmy possum, and its snout is about two-thirds as long as the rest of its head. The tongue is bristled at the tip, and it is so long it can be pushed out almost half the length of the animal's body. Honey sucker is another name for it.

Honey suckers live in southwestern Australia where blossoms of some kind, usually on trees or shrubs, can be found all year. The animals feed by thrusting the long tongue through tubelike lips deep into a flower. There it becomes covered with sticky nectar and pollen. When the tongue is pulled inside again, ridges on the roof of the mouth wipe food from it. Then the brush-tipped structure is darted back into the flower.

Honey suckers have short, coarse fur that is a beautiful,

rich brown on the top side and creamy white underneath. A dark stripe runs down the back, with a lighter streak on either side. The long tail is whiplike and hairless, and can be curled at the tip.

These small animals are active and can climb rapidly. Each one usually lives alone but when a plant is covered with blossoms a crowd of the creatures may gather, all feed-

Honey possums feed on nectar and pollen from flowers. In Australia this animal is called a noolbenger, a name adapted from that used by the aborigines.

ing at the same time. Sometimes one will hang by its tail in order to reach out for a flower and pull it closer. Then the animal is up and off again, hunting for more nectar.

Animals and Man

These friendly little noolbengers, like many other animals, have few enemies other than man. They live in woodlands and heaths, building nests and raising families. No creature is more harmless.

Yet, year after year, their homes are destroyed as man rips up the countryside and turns woodlands into pastures and fields and great, snaking highways. In the southwestern region of Western Australia alone, about 400,000 hectares of new land is brought into production each year. This is equal to about four billion square meters, or a million acres.

Nations need food, but adopting short-sighted measures for quick results will, in the long run, do more harm than good. In all countries huge changes are threatening many animals with extinction. Unless far-seeing plans for future use of land are made soon, including reserves and refuges for wildlife, not only small honey possums but many other defenseless creatures will die.

The key question facing all of us now is: If nature's balance is so upset that most animals die, can the human race itself live?

8

A Country's Symbol: The Big Leaper

Few animals of Australia are as well known as the great, leaping kangaroos, yet not many people realize that true kangaroos can be almost any size. Some species, such as the big red or the gray forester, are taller than a man. Others, when fully grown, are not as large as a rabbit, and there are all sizes in between.

The large kangaroo is one of Australia's national emblems. Pictures of the animals appear on stamps, and their figures decorate the tails of great airliners. The half dollar shows the Commonwealth's coat of arms being held on one side by a kangaroo and on the other by a huge native bird, the emu, that looks much like an ostrich. One reason these two animals were chosen is that neither can go backward.

Macropods

Members of the kangaroo family are known as macropods, a word which means "big feet." The hind legs of the animals are large and so strong they seem to be made of steel springs. As in phalangers and several other marsupials, the second and third toes are joined to form a twin toe that is used in combing the fur. The big fourth toe has a long pad under it that reaches from the toe tip to the heel. The front limbs of macropods are usually quite small, and the little paws are used as hands.

A kangaroo uses its big, muscular tail as a prop when it

A joey, or young kangaroo, is seen at left, top to bottom, in the pouch soon after birth; at an older stage; and leaping into the pouch when alarmed. Above, a joey looks out on the world from the safety of the pouch and can even feed on grass while its mother does the same.

is sitting up, so that its body forms a sort of three-legged stool. During a fight the tail is used as a brace. The animal can lift one or both feet at the same time, and rip open with its big claw the underside of a creature that is jumping to attack it.

The tail also acts as a balancer when the kangaroo takes off over the countryside in great jumps. One leap can carry the animal three or four times its own length, and it can easily sail over a fence its own height. If hard pressed, it can vault over objects twice that high. Kangaroos have been clocked at speeds up to 40 or 50 kilometers an hour (about 25 to 30 miles), but they cannot keep this up for any length of time.

A macropod's teeth are suited for grazing or for browsing on small twigs. Two front teeth in the lower jaw jut forward like scissor blades. The upper teeth are smaller, and ridges just behind them help the animals hold grass in place as they nip it off. Canines are often absent, but the first two pairs of cheek teeth are somewhat bladelike. The back teeth are broad and have grinding ridges much like those on teeth of cows or horses.

"Big Red"

The plains kangaroo, or "big red" *(Megaleia rufa)*, grows to be taller than any other kangaroo, but it is not always the heaviest. It lives on grasslands where water holes are often far apart. The fur of the male is a reddish color, much like that of deer, but the smaller female is bluish gray. During mating season the chest and back of a male become a bright red because of gland secretions. This substance is powdery and rubs off easily.

Reproduction in most macropods depends a great deal on certain conditions in the environment, such as the amount of food or water. Young kangaroos of pouch age are known as "joeys" and in plains kangaroos are born about 33 days after mating has taken place. Four nipples are in the pouch, but usually only one infant is born at a time.

The red mother mates again a day or two after giving birth, but the second embryo remains in the uterus as a kind of reserve. It grows to be a ball of about 100 cells, then goes into a resting period. No further development occurs as long as another young one is in the pouch. This kind of delayed reproduction is found in most macropods, but not in other marsupial families, except for one of the pygmy possums.

The joey first starts making small noises when it is about two and one-half months old. Four months later it begins leaving the pouch for short periods, but dives back in again, headfirst, if danger threatens. Then it turns around and peers out again to see if all is well. An eight-month-old joey is too large to crawl into the pouch and at this age is feeding on grass, but it will stand in front of its mother and poke its head into the pouch to nurse as long as she will allow it to do so.

Nearly four months after the older joey begins leaving the pouch for short periods, the reserve embryo begins to develop again. It is born a day or two after the older joey leaves the pouch for good. This newborn infant clamps onto one of the three small nipples, which are still a normal size. The fourth nipple, which is being used by the older juvenile, is long and is nearly as large as the new infant. For the next few months, while both youngsters are nursing, the mother produces two kinds of milk.

In all mammals, whether they are placentals or marsupials, the make-up of the milk changes as the infant grows. At first there is much sugar. Later, when the infant is older, less sugar is produced and more fat and protein appear. Scientists have been astonished to find that mother kangaroos can make two types of milk at the same time.

Milk received by a year-old joey has about three times as much fat in it as milk from the small nipple used by the newborn. There is not only more protein in the older youngster's milk, but new kinds of protein are produced that are specialized for its body needs. As yet no one is able to understand how the glands of the mother are able to control this difference.

Seasons of drought sometimes occur on the plains when water is scarce and the food is of poor quality. In conditions of this kind the mother seems not to be able to produce rich milk for the older joey, and it soon dies. Then the little reserve infant is born and takes its place.

If conditions grow worse, all the young die. No more are born until after rains come and green shoots start to grow. With good food and enough water the females soon become ready to breed again. In good seasons, about three young joeys are produced in four years.

Small groups of kangaroos often band together, forming a larger group—a "mob"—as they roam from place to place. One individual, usually a male, bosses the others. The animals feed at night and are likely to spend most of the day resting in some shady spot.

Sometimes a stranger comes up wanting to share another's patch of shade or drink at the same water hole. The new arrival is likely to be driven off by loud snorts, but if he insists on staying the animals may fight, cuffing each other with their forelimbs and holding their heads out of the way. Or the two may simply sniff noses, then pay no more attention to each other.

Gray Foresters

The great gray, or forester kangaroo *(Macropus giganteus)*, lives in eastern Australia and in Tasmania where rainfall is more regular. Great grays are not often as tall as the reds, but they are usually heavier and more powerful. Both males and females have silvery-gray fur and a somewhat hairy muzzle.

Habits of the reds and grays are much alike. Both live in small groups, although they may travel as part of a mob, and both are more likely to graze on the spiky grass known as spinifex than to browse on tender shoots from woody trees or shrubs. Both can make long leaps, but the great gray usually holds its forequarters low and its head high. As it springs along, its tail is curved upward, and sometimes goes up and down as if it were a pump handle.

Great grays in captivity seem to be more gentle and trusting than other kangaroos, and also more nervous. One female in the Zurich Zoo was so tame that she allowed keepers to examine her pouch at any time. She even cooperated when they took X-ray pictures of her young joey jumping headfirst into her pouch then turning around to look out over the pouch's rim.

Great grays do not mate soon after a new infant is born as most macropods do. No reserve embryo is carried until the young one in the pouch is well developed and has been nursing for some time. The joey does not leave the pouch until it is eight or nine months old. Then it hops in

The great gray kangaroo is larger than the red.

and out for another two or three months, and may nurse until it is a year and a half old.

Wallaroos

A third kind of macropod is nearly as large as the red and the gray kangaroos, and is often mistaken for them. This is the wallaroo, or euro *(Macropus robustus)*, sometimes called the hill kangaroo. It is found in rocky outcrops and gullies all over Australia. The fur of the euro is long and coarse, and that of the males is usually quite dark.

Wallaroos reproduce much as red kangaroos do. A young embryo is held in reserve and can take the place of an older joey during a dry season. These two species can be crossed, but the hybrid joey that is born is sterile and can never have offspring of its own. This occurs with certain other animals also—for example a mule, which is the result of a cross between a mare and a donkey.

Euros can live on poor land with almost no water if there are boulders and caves where the animals can escape the heat of the sun. Research has shown that a kangaroo, if given a choice, will eat coarse grass rather than pasture plants. Weight for weight, a kangaroo eats about as much as a sheep does, but makes better use of its food. A kangaroo is 52 per cent edible meat but a sheep is only 27 per cent.

Kangaroo Slaughter

When sheepherders moved on the Australian plains they burned off the land, planted pasture grass, and dug wells. Then they often overstocked the pastures and tried to raise more sheep than the land would support. Soon the pastures were overgrazed, spiky grass grew again, and many sheep died. They could not live on this poor food.

Such conditions were ideal for kangaroos. They began to increase in numbers, year by year. Angry farmers complained that the animals threatened their livelihood and began to poison them.

Soon shooting of the animals began. The hides were sold for everything from shoes to Scotch bagpipes. Kangaroo hams were sold to Oriental countries and tons of meat and other parts were used as pet food. Hunters joined forces. They used cars to chase the animals, and spotlights to scare the creatures into standing still so they could be shot more easily. It was possible to make up to $70 a day in this way. Hundreds of thousands were slaughtered.

Finally the states of Australia, one by one, stepped in and began to make rules and regulations similar to our own game laws. It is now illegal to send hides or meat out of the country without permission. Although kangaroos are protected by law, poachers still kill them by the thousands and sell them to dishonest buyers.

Early in 1975 both the big red and the gray forester were put on the endangered list. This means that every year more are killed than are being born. Unless laws against poaching are enforced soon, the world will lose forever two of its most unusual creatures.

The Prettyface Wallaby

Great kangaroos have feet over 25 centimeters (about 10 inches) long. The common macropods that are medium-sized and have smaller feet are known as wallabies. Many of these animals are quite graceful and are often richly colored.

One of the most striking of the larger wallabies is the prettyface *(Macropus parryi)*. It is a pale gray color with marked white stripes on its face and a white streak that goes from hip to hip under its tail. Other names given this animal are gray face, blue flier, and whiptail. The last name refers to the long, slender tail which is usually held in an upright curve when the animal hops.

Whiptails live in grasslands and woodlands in eastern and mideastern Australia, usually near hills or slopes. They are the most social of all the macropods. Great grays and red kangaroos live in loose mobs, but the whiptails form large herds that contain all ages and sexes, and stay together most of the year.

During poor or dry winters, two or more mobs may feed together peacefully. The males rarely fight unless one is trying to drive off other males from a female that is ready to mate. Then the animals cuff and box, but neither one is harmed. The territory covered by a big mob is usually about 100 hectares (about 250 acres).

Swamp Wallabies

Eastern Australia is also the home of the black-tailed wallaby *(Wallabia bicolor)*. The animals live in dense thickets, especially in gullies and other places where the ground stays wet much of the time. This has earned for the creatures the name "swamp wallaby."

The color of the fur changes according to locale. Northern swamp wallabies have brownish gray heads and backs, and orange undersides. In the south the black-tailed wallaby is darker all over; even its belly is brownish red. These animals hop with the head bent down and the whole body hunched over.

In an interesting experiment done a few years ago a small swamp wallaby was placed in the pouch of a red kangaroo, which became its foster mother. It stayed in the pouch as long as swamp wallabies usually do, or about a month longer than red kangaroo infants. This wallaby grew faster than normal, and became adult at an earlier age than wallabies do ordinarily.

What made such a difference? Was it the kind of milk? Or the amount? As yet no one knows.

Pelsaert's Discovery

The small scrub wallaby, or tammar *(Macropus eugenii),* is probably the first of the kangaroo family ever seen and written about. When the Dutch captain Pelsaert was shipwrecked on the western coast of Australia in 1629 he reported, after his rescue, that he had seen a large animal that hopped on its hind legs. Furthermore, this strange creature carried a young one, attached by its mouth, in a pocket of skin on its belly.

From the rest of his description it seems to be the tammar wallaby that Pelsaert saw. No one paid much attention to his wild story, and it was forgotten for a century and a half. Not until Captain John Cook landed in Australia and brought back some actual specimens did people believe that an animal such as this really existed.

Tammars live in dry, dense thickets in southwestern Australia and on some offshore islands. The animal is chunky and has grayish brown fur with no distinct face stripes. Sometimes the shoulders are reddish brown.

The tammar is remarkable because of its ability to live under very dry conditions. Dr. A. R. Main, of the University of Perth, kept several of the animals in cages for over 30 days and gave them only sea water and dry food. The tammars remained in good health and even gained weight.

Many islands where tammars live have little rainfall and no other fresh water of any kind. The animals have been seen drinking sea water. Their kidneys are specialized for this sort of life, and most of the water taken into the body is re-absorbed from the urine before it is passed out.

Small Pademelons

A number of the smaller scrub wallabies are known as pademelons. This is an English mispronunciation of the native Australian name for the animals. Pademelons are curious creatures and sometimes allow one to come quite close before bounding away. Like rabbits, they have the habit of thumping the ground with their hind feet, probably to signal a warning to others nearby.

One of the more common of these creatures is the Tasmanian pademelon, or red-bellied wallaby *(Thylogale billardierii)*. It is found in low scrub on several islands and on the southeastern coast of the mainland, as well as in Tasmania. The animal is dark brown, with small, round ears and a short tail. It is sometimes caught for its soft, thick fur.

Scrub wallabies live in tunnel-like runways, and eat leaves and shoots as well as grass. Because of its eating habits the Tasmanian wallaby interferes somewhat with keeping the forests in good condition, but it is not yet a real problem.

First to Be Studied

The quokka *(Setonix brachyurus)* is one of the smallest of the wallabies. It is important because it was the first Australian marsupial to be used for scientific studies in physiology and behavior. Until the 1930s it could be found in most of the wetter regions of the mainland, but only a few colonies are now left. Most of the animals live on islands off the west coast where summers are dry and hot.

Quokkas are stocky little creatures not much larger than a cat. The fur is a coarse brownish-gray, with a reddish tinge on the head and chest. Ears are small and rounded, and set high on the head.

The hind feet of a quokka are short, but the animal hops as a kangaroo does. The tail is short and is not used as a prop, but when the animal rests it sometimes brings its tail forward between the hind legs and sits on it. Then the head is tucked down until the creature looks like a round, furry ball. Quokkas, like pademelons, eat grasses and other plants, and make runways through the vegetation.

Males sometimes fight each other for a few seconds, biting and kicking, but most arguments are settled by the animals' sniffing or cuffing each other, or by one running away. One male usually "bosses" the others. A quokka will move away from a feeding pan, or from a female, if a more dominant male comes along.

Most quokkas endure the hot weather by gathering at "soaks" when the temperature rises. These are places around lakes that are drying up, where the animals can dig and find water. The body temperature is kept almost normal even when the outside air reaches 44° C. (over 110° F). The animals pant, sweat from the paws and forearms, and lick their arms and spread the saliva over the body. On parts of the islands where no soaks are found the animals eat certain plants which contain stored water.

Scientists and Quokkas

About a century and a half ago studies were made on quokkas at the London Zoo. One after another the animals were seen to bring up part of their food, chew it again, then swallow it, much as cows do. Later the stomachs of the animals were found to have four parts, and to be similar in several ways to those of placental cud-chewers. Both groups of animals also nip off grass stems, then pass the food back to be ground by the cheek molar teeth.

No mammal is truly able to digest cellulose, the tough, woody part of plants. Studies made about the middle of this century showed that most macropods, like our more familiar cud-chewers, have special bacteria in the forepart of the stomach whose chemical enzymes work on the cellulose. When the tough fibers are broken up the animals can then finish digesting the plants.

Delayed birth was also first studied in quokkas. In these animals the early stages of "reserve" embryos are much like those found in large kangaroos, but later stages are different. In both groups the newly fertilized egg grows to be a ball of about 100 cells, then develops no further if a young one is in the pouch.

When a young quokka is nearly ready to leave the pouch, the resting ball of cells in the uterus is broken down and destroyed. If some accident happens to the joey before it is old enough to leave the pouch, the ball of cells will continue development and take its place, but if the older youngster finishes normal development the reserve is no longer needed. It dies, and another embryo is not produced until the mating season the next year.

Hare Wallabies

One group of small macropods is known as hare wallabies, probably because the animals have rabbit-sized bodies with long, pointed, rabbit-like ears. All of these creatures have orange rings of hair about their eyes, but in one species

The hare wallaby is no longer seen on the Australian mainland. It is capable of high speeds.

this shows up so well that the common name is spectacled hare wallaby *(Lagorchestes conspicillatus)*. This scientific name means "conspicuous dancing hare."

The fur of the spectacled hare wallaby is yellowish or brownish gray above and lighter underneath. White markings are on its hips. Like other members of this group it lives in grasslands and makes its nests under bunches of grass or bushes.

These animals are so built that they can achieve great speeds. They have great hind limbs with narrow feet, and can outrun dogs for short distances. If one is disturbed it comes out of its nest with a rush and bounds off. It is no trick at all for one to jump over a man's head.

The animals are no longer common. Some species have not been seen on the mainland since early in this century, but several populations live on islands off the western coast. Such islands make good natural laboratories, and scientists are now studying species which live there.

Rock Wallabies

Much of Australia, especially inland from the eastern and western coasts, has ranges of hills and mountains. Where boulders, caves, or rocky outcroppings occur, rock wallabies can usually be found. The animals are a little larger than hare wallabies, and are as surefooted as mountain goats.

Both tail and feet of rock wallabies are especially adapted for the sort of life they lead. The tail is well furred, long and thin. A tuft of hair at the end helps it act as a balancer. The nonskid soles of the feet are thickly padded and covered with rough skin. Stiff hairs edge the soles.

The common brush tailed rock wallaby *(Petrogale*

The brush-tailed rock wallaby is a cave-dweller. These marsupials live in groups.

penicillata) is found over most of Australia. It is a rich dark color with white cheek stripes and a black patch in each armpit and on each ear. A pale strip behind the arm patch, and white margins on the ears, make these dark spots show even more clearly.

Rock wallabies live in groups in caves. They rest by day and come out in the evenings to feed on grass, or the bark and roots of various trees. The animals can climb anywhere, leaping from ledge to ledge up a cliff face, or jumping across deep chasms three or four meters (ten or twelve feet) wide. They often seem to chase each other for fun, and rocks become smooth where they are active a great deal.

These graceful creatures move awkwardly on flat ground, and rarely are found on it. If danger threatens they give a loud, thumping signal with the hind feet, as pademelons do, then vanish into the nearest crack in the rocks.

The Nail-Tails

Animals in one odd little group of macropods have a dark, horny nail somewhat like a fingernail—without any known function—hidden in the fur at the tip of the long tail. For this reason they are known as nail-tailed wallabies. Only three species live in Australia. Most are now rare except in the northern part of the continent where the karrabul, or northern nail-tail *(Onychogalea unguifera)*, is still rather common.

This fox-sized animal is the largest of the three. All have a square muzzle and a dark stripe that runs down the back and on the long, whiplike tail. White markings are on the cheeks, hips, and shoulders but these do not show well on the karrabul because of the light sandy color of its fur. Nail-tails are sometimes called organ grinders because they have the curious habit of holding their arms out to the side and moving them around in a circle as they hop.

Nail-tails live in grasslands or scrub, but are more likely to dig for roots of coarse grasses than to eat the leaves or tops. Like other wallabies they have long, strong hind legs and feet. If one is disturbed it utters a soft "Uh-uh-uh" and bounds off quickly.

Tree Kangaroos

The wallabies that live in trees are probably least like other macropods. Two species are found in northeastern Australia, but several others live in New Guinea. Although not rare, the animals are not well known because they live in mountainous areas or rain forests far from cities or farms. Their main foods are leaves and fruit.

Scientists think that this branch of the macropod family probably migrated to New Guinea long ago, then developed the habit of living in trees during the thousands of years that the big island was cut off from the Australian mainland by rising ocean waters. Later, when Torres Straits could be used as a pathway again, some of the animals returned to Australia. Although this land bridge is now underwater, and has been for thousands of years, the two groups of tree kangaroos are still alike in many ways.

Tree kangaroos are a little larger than nail-tails, and have powerful forearms nearly as long as the hind legs. Rough, cushion-like pads on the broad hind feet make it easy for the animals to leap from tree to tree, or from a high branch to the ground. The long, furred tail is used as a balancing organ.

Some of the New Guinea species are brightly colored. Matschie's tree kangaroo *(Dendrolagus matschiei)* has a reddish back with a bright yellow face, paws, and tail. The underside is also yellow, and two lines the same color run along its spine. The animals sleep huddled in tree crotches, with heads tucked down beneath the paws. A "cowlick," or natural whorl of hair, on the neck allows the water to be shed in all directions when tropical rainstorms beat down.

The colorful Matschie's tree kangaroo has bright yellow face, tail, and paws, and a reddish back.

A young Matschie's tree kangaroo was born in Columbia Zoological Park near Columbia, South Carolina, in August, 1973. This new park was first opened to the public in April, 1974, and the infant tree kangaroo was the first young animal of any kind born there.

Rat-Sized Potoroos

The smallest macropods are usually known as rat kangaroos because most are rat-sized or slightly larger. One group, the potoroos, lives in forests or woodlands, or areas where there is much rainfall. At first glance these animals look much like small bandicoots.

Probably best known is the long-nosed potoroo *(Potorous tridactylus)*. It is still found in several places on the mainland but, like other small marsupials that are preyed upon by meat-eaters, it is becoming less common. During the summer potoroos feed on both plants and insects, but they depend mainly on fungi, mosses, and similar plants the rest of the year. Some potoroos reach an age of six or seven years.

Dry-Country Boodies

Rat kangaroos that live in dry country are known as bettongs. One species, the boodie *(Bettongia lesueur)*, has been studied a good deal at the Australian National University and in the Division of Wildlife Research at Canberra.

Most bettongs make grass nests above the ground, but the boodie digs burrows. It collects straw with its mouth and piles it on the ground, then uses the hind feet to tramp it down and make a bundle. This is carried to the nest in the downward-curved tip of the tail. The young are born about three weeks after the parents mate. The mother then carries

Top to bottom, the musky rat kangaroo, the boodie, and the long-nosed potoroo

a reserve embryo for about four months until the joey is ready to leave the pouch.

Boodies live in groups. Usually these are made up of one male and two or three females. Males will fight to defend the females, but they never use their tails as props while they cuff and box. Instead, they lie on their sides and kick each other, lashing out with their hind feet.

Boodies were common on the mainland until the 1930s, but now are quite rare. Populations of the animals are still common on some of the islands off the western coast.

A Rather Different Kangaroo

The musky rat kangaroo *(Hypsiprymnodon moschatus)* is so different from other macropods that it is sometimes placed in a subfamily by itself. Its tail is naked and scaly, and it has five toes instead of four on its hind feet. Its fourth toe is longest, however, and the second and third toes form a grooming claw. It hops about as kangaroos do.

This small, musky-smelling creature is found only in the rain forests of northeastern Queensland. In spite of its smell the animal is rather attractive, for its velvety fur is a rich, reddish brown and the underside is paler with an orange cast. Unlike most macropods, musky rat kangaroos are more likely to have twins than one joey, and they seem to be as fond of insects and worms as of roots and berries.

A Distinctive Family

Members of the kangaroo family have reached the highest degree of development found among marsupials. If Australia had not been separated from other continents for

millions of years, such specialized creatures would probably never have appeared. Much of the fascination of kangaroos and other marsupials lies in the fact that they are different in so many ways from the other animals we know.

Yet, unusual as they are, marsupials are not the only creatures which make Australia one of the most remarkable continents on earth.

9

A Very Strange Continent

Australia has often been called "nature's museum" with good reason. No other land has great, plant-eating creatures that can leap across a busy highway in one jump; pouched animals that glide through the air; or cuddly "teddy bears" that must eat leaves of a certain kind of gum tree or die. Even the fierce, meat-eating Tasmanian devils can become playful, affectionate pets. Australia's animals seem to have walked from the pages of a science fiction book.

But even more unusual creatures live here. They, too, are mammals, but are neither marsupials nor placentals. They are monotremes, an odd group of mammals without relatives anywhere else in the world. Even fossil remains are almost unknown. To get a better idea of marsupials, and mammals in general, we should have a look at them.

Monotremes are so much like reptiles that zoologists wondered for years whether or not they were really mammals. The creatures laid eggs, as reptiles and birds do, but the mothers nourished their newborn infants with milk. Hair, instead of scales or feathers, covered their bodies.

All sorts of questions were raised. Were these animals some peculiar kind of reptile that had developed hair instead of scales? Were they once mammals that had lost some of their characteristics and become like reptiles again? Or were they some sort of strange combination that formed a living link between the reptile and mammal groups? No one was sure about the answers.

A Patchwork Puzzle

One of these odd creatures is the duckbill platypus *(Ornithorhynchus anatinus)*, a flat-bodied, compact little animal not much larger than a cat. It has a beaver-like tail, glossy brown or black fur that is paler on the underside, and a soft, leathery muzzle shaped like a duck's bill. On the top side of the snout are two nostrils. A short flap of skin projects backward, covering a streak of fur across the front of the head.

The platypus has short, stout legs with broad, webbed feet, and claws on every toe. The front feet are used for swimming, but the hind legs trail behind, keeping the body on an even keel. On each side of the head is a fold of skin that keeps water from the animal's eyes and the openings to its ears. The web is thick on the front feet, and juts out beyond the claws. It is a help in swimming, but when the platypus waddles about on land the extra skin has to be folded back under its palms, out of the way.

These animals were first discovered in 1796, and about two years later a dried skin was sent to the British Museum in London. Dr. George Shaw, of the Natural History section, thought he had received a fake. Sailors often brought home stuffed "Chinese mermaids" made of monkey skin carefully sewed to a fish's tail. To prove the queer animal was a joke, and not real, Dr. Shaw tried to pry off the bill. The same dried skin is in the British Museum, and the marks of Dr. Shaw's scissors can still be seen.

Underwater Mammals

Platypuses are common in fresh-water streams and lakes of eastern Australia and Tasmania. They live in long, winding burrows built into the banks. Openings are a short distance above water level and are sometimes flooded, but the rear ends of the long tunnels are too high for the water to reach them. Both males and females live here except during the breeding season.

Feeding time is in the early morning or late afternoon. The platypus heads for water, then ducks its head as it swims and noses for food in the mud and gravel on the bottom. Thousands of tiny nerve endings, sensitive to touch, cover the broad snout.

An adult platypus has horny ridges on each jaw instead of teeth. These are sharp at the front, but flat and strong at the back. As the animal sculls along, head downward, it scoops up crayfish, tadpoles, and other small water animals, storing them in its cheek pouches. Every few minutes it comes to the surface, crushes and swallows its prey, then dives into the water again. Each animal eats about half its weight in food each day.

In the spring, at mating time, the female builds a special breeding burrow just large enough for her to squeeze through. No male is allowed to enter. At the far end is a nest lined with wet grass and leaves. The female carries in these materials with her tail, folding it forward under her body between the hind legs.

Top to bottom (rt.): the platypus lays only two or three eggs; young platypuses get their nourishment by licking and sucking milk from the mother's hair; food is found in mud and gravel at the bottom of a stream or lake.

Mating takes place in the water. The male courts the female by grasping and nibbling her tail with his bill while the two swim slowly in circles. About two weeks after mating the female goes to her nesting chamber to lay the eggs. As she wriggles through the tunnel she blocks it tightly in two or three places, tamping the dirt in firmly with her tail.

Two or three eggs are laid. They have leathery shells and are covered with a thin film that helps them stick together. Each one is rounded and is a little less than two centimeters long (about 5/8 inch). The mother incubates the eggs by curling her body around them, or by holding them against her belly with her tail, for she has no pouch.

Newborn Duckbills

Ten to twelve days later the eggs hatch. The small creatures are blind and naked, and not much larger than infant opossums. Platypuses have no nipples, but thick, yellowish milk oozes from milk glands on the mother's belly. The young ones lick and suck the milk from the hair. Before long a few teeth start to grow, but they soon disappear.

The babies are not weaned until they are about four months old. During this time the mother must eat a great deal so she can produce plenty of milk. Each time she leaves the burrow to hunt food, or goes back in, she unplugs each place she has built a block, then presses the earth back in place again.

Poison Spurs

All young platypuses have a hollow growth on the inside of each ankle. It soon disappears in females, but in males it grows to be a horny spur that is connected to poison glands in the thigh. When a platypus strikes with his spurs, he drives his hind feet together and stabs anything caught between them.

No one knows exactly what these spinelike structures are used for. Possibly they serve some function during mating, or are an aid in fighting enemies. David Fleay, who has raised platypuses in captivity, saw one animal jab a frog which it later ate. If the poison is accidentally injected into a person, it is not dangerous but is quite painful.

Young platypuses follow their mother at first but soon learn to find their own food after they leave the nest. Two years later they are ready to mate and produce families of their own. No one knows just how long platypuses live, but a male captured wild, and placed in the MacKenzie Sanctuary at Healesville, Victoria, had been there 17 years when it died.

A Group Set Apart

Monotremes differ from other mammals in several ways. The word "monotreme" means "one hole" and is used because these animals have only one opening at the rear of the body. Marsupial and placental mammals have an anus through which wastes from the intestines pass, and one or two other openings for urine and for sperm and eggs. In monotremes all these tubes open into the last part of the large intestine, so everything must leave the body through the one outlet. Reptiles also have but one body opening at the rear.

Monotremes are like reptiles in several other ways. Animals in both groups lay leathery-shelled eggs and incubate them. The shoulder girdles and some bones of the back and ribs in monotremes are much like those of reptiles. Some of the eye structures are reptile-like, too.

On the other hand, some of the most important features of monotremes are found only in mammals. Neither fur nor milk glands occur in any other groups. Monotremes have one bone in each lower jaw, and three bones in the middle ear, as other mammals do, but reptiles and all other vertebrates have several bones in the jaw and only one in the ear.

Monotremes also have a pair of marsupial bones in front of the hips, like those of marsupials. No one has discovered what these bones are used for in either group. Perhaps they are a "carry-over" from some ancient ancestor and have no value for animals that live today.

A Spiny Monotreme

Only one other kind of monotreme is known. This is the echidna, or spiny anteater. One species, the short-nosed spiny anteater *(Tachyglossus aculeatus)*, is widespread over Australia and Tasmania. Two or three long-nosed forms live in New Guinea and some other islands nearby. These used to be common in Australia, too, for many fossils of this kind have been found on the mainland.

The short-nosed echidna is a stocky, heavy-set animal with short legs and a naked, stubby tail. Its body is larger than that of the platypus and is covered with long, sharp spines that are usually yellow at the tip and black at the base. Coarse brown hairs grow between the spines. Thick bristles instead of spines cover the underparts.

The animal's head seems to be set directly on its squat body. No outer ears show, but two bright eyes can be seen at the base of the slender, beaklike snout. A small mouth at the tip of the snout is just large enough for a wormlike,

The echidna, one of the monotremes, is a spiny animal that lives on ants.

sticky tongue to flick in and out quickly. The scientific name of this animal means "sharp, swift tongue."

Spiny anteaters are powerful diggers. The claws are heavy and curved and the animal, if disturbed, can dig almost straight downward at great speed. One echidna, digging in hard soil, disappeared in nine minutes. When the animal is well hidden, or braced by rocks, it is almost impossible for another creature to dislodge it.

Spurs grow on the hind legs, as in playtpuses, but are not attached to poison glands. The second toe on each hind foot is extra long and is used by the animal to scratch between its spines. An echidna can run swiftly, but it waddles when it walks because the toes on its back feet are turned outward and backward.

The favorite food of echidnas is termites and some species of ants. The animals search actively for the insects, sniffing, pawing, and poking. They investigate everything. Sometimes great mounds of earth are torn open, or an animal in captivity may dig under a heavy food trough and tilt it over.

Pouches for Eggs

Even though spiny anteaters spend much time digging, they do not make nesting burrows. Instead, the female develops a temporary pouch and lays one or two eggs in it. No one yet has seen how this is done. The egg is too large for the animal's mouth and too delicate to be handled by the great, digging claws.

Many zoologists think the mother curls herself enough to lay the egg directly in the pouch. This would be possible, because an echidna can roll up so tightly it looks like a spine-studded ball. The egg hatches in about a week and a half, and the new infant laps milk that oozes out on its mother's fur, just as a platypus does. In six to eight weeks, when spines begin to develop, the mother places her prickly baby in a den, but comes back regularly to feed it.

Dr. M. Griffiths, of the Division of Wildlife Research, watched the feeding process. The mother used her beak gently to nudge the baby until he lay between her forelegs, then pushed him back under her body. He grabbed her underfur with his paws and sucked for about half an hour, hanging upside down.

Echidna Hibernation

Monotremes are able to control their body temperature to some extent, but usually the temperature is lower than that of another mammal the same size. In Tasmania and in some parts of southeastern Australia the spiny anteaters sometimes go into a state of hibernation. During this time all body processes go on very slowly. This state may last a week or more at a time.

Most animals that enter this condition do so during cold weather after they have stored a good deal of fat. Echidnas are curious in that low temperature or fat storage does not seem to be as important as starvation. An animal that weighs over four kilograms (a little less than nine pounds) will not hibernate.

Monotreme Fossils

Almost nothing is known about the most ancient ancestors of today's monotremes. A few peculiar teeth, about ten to twenty million years old, have been found near Alice Springs in central Australia. It is possible that these belonged to early monotremes, but it is difficult to be sure, since adult monotremes today have no teeth at all.

The only known fossils of monotremes are fairly recent ones from the Pleistocene Ice Ages. In Australia, as in other continents all over the world, many giant forms lived during this time. Some of the platypuses living then had bodies as large as those of the wolves that live in North America today. Fossils of giant echidnas have also been found. Other monotreme fossils are smaller, and the animals were probably the same species as those living today.

Several guesses or theories have been published about the early ancestors of monotremes. Most scientists today think that these animals are not very closely related to marsupials, and are probably descended from a different kind of mammal-like reptile. Perhaps other fossils will be discovered someday that will tell us more about the history of this curious group.

Marsupial Fossils

Although no one knows just how long marsupials have been in Australia, most of the fossils found so far date back to the Miocene or later subperiods. A few, discovered recently, seem to have been laid down during Oligocene times. This would make them about 30 to 35 million years old.

Because these early fossils show that many different kinds of marsupials already lived in Australia at that time, some scientists think the animals may have come to Australia at a much earlier date, perhaps 50 or 60 million years ago. From what we know of the history of the shifting continents, this would be after South America became separated from other lands at its southern tip, but before Australia broke away from Antarctica and started its journey northward.

During Pleistocene times many marsupial giants developed. Nearly every family had at least one giant species, and some had several. Immense kangaroos measured 3.5 meters, which is nearly twice the size of the big macropods of today. The sthenurids, or strong-tailed kangaroos, were even larger. They had thick, heavy tails and shortened muzzles, and browsed on tender tree shoots. This subfamily of kangaroos died out many thousands of years ago.

One of the oddest giants was Thylacoleo, the marsupial lion. It was the size of a true lion, and shapes of certain bones show it was related to the phalangers. Yet its teeth were like those of no other animal. Some in front were dagger-like and the cheek teeth were as sharp and able to shear flesh as those of meat-eaters.

For a long time Thylacoleo was thought to live on fruits, but an entire skeleton of the animal, recently discovered, may change the picture. The pouched lion had the heavy limbs and clawed feet of a carnivore. It seems this odd creature was, after all, a beast of prey, in spite of its many plant-eating relatives.

The largest marsupial fossils ever found are those of Diprotodon, a great, waddling creature as large and clumsy as a rhinoceros. It had gnawing teeth and looked much like a huge wombat, but was not closely related to it. These animals lived in large herds, grazing on the rich grass that once grew in what is now desert country. All of them are now extinct.

Diprotodon, now extinct, was a giant among marsupials, being about the size of a rhinoceros.

Even the parasites of some of the giant animals were oversized. The fossil of a king-sized bandicoot that lived 120 million years ago has been found east of Melbourne. With it were fossil fleas that measured nearly a centimeter (about one-third inch) in length.

Early Placentals

A few placental mammals have also lived in Australia for millions of years. Many people do not realize that some of these are "native" Australians, although fossils show they arrived long after marsupials had spread to all parts of the continent. Both rodents and bats were among the later "natives."

True rodents are gnawing animals. They are found everywhere except in the sea. Such creatures as squirrels, gophers, rats, mice, beavers, and porcupines belong to this group. Only a few species of rats and mice reached Australia, but fossils of some rodents are up to 20 million years old.

By the time the rodents arrived, Australia had drifted far to the north, toward Asia. Many islands now lay between these two great land masses. No one knows how the gnawing creatures reached the southern continent, but they make such good "stowaways" on ships and sailing vessels that over the centuries they may have caught rides on rafts of floating logs and gone from island to island. In this way they could have finally reached Australia.

Native Rodents

Four groups of Australian native rats and mice have developed. The bush-dwellers are widespread, but are so secretive they are rarely seen. Water rats, a second group, live on New Guinea as well as the mainland. One of the most interesting of these creatures is the beaver rat. Except for the platypus, it is the only fresh-water mammal that lives in Australia.

A third group, the mosaic-tailed rats, have a peculiar pattern of scales on their tails. These animals live mainly in trees, and are also common on New Guinea and other islands nearby. Farmers dislike the creatures because some of them damage sugar cane and coconut crops.

Members of the fourth group are more truly Australian than any of the others, because they have no close relatives in other lands. Stick-nest rats, rock rats, and certain tree rats belong here, as do the native mice. These are soft-furred little animals with large ears that live in all kinds of places. Forests, deserts, mountains, woodlands, and grasslands are all home to some creature in this group.

Australian Bats

Bats have a long history in other parts of the world. Some of the earliest fossils of these mammals are 50 million years old, yet the animals were already specialized enough that they could truly fly. Their wings are thin membranes, stretched wide by long, thin fingers that form stiff ribs. No one knows when bats first arrived in Australia, since no early fossils have been found, but reaching the continent was apparently no problem for animals that could fly. Earliest bat fossils in Australia date back to the Pliocene, but it is likely that most of the bats came down from the north during the middle or late Pleistocene subperiod, for all bat families found on the continent occur also in Asia or the East Indies.

Bats are often feared but should not be, for most of

Bats reached Australia at least as far back as the late Pliocene epoch. This is the flying fox, a fruit eater. Other bats are seen hanging upside down from the tree in the background.

them are helpful. Small ones usually feed on insects, and can eat one-third the weight of their bodies in an hour. Large bats, commonly called flying foxes, live on fruit and are at times a pest to fruit farmers. None of the small vampire bats, or bloodsuckers, live in Australia.

Boat Travelers

Two other groups of mammals have made Australia their home for thousands of years. Australian aborigines are among the oldest peoples on earth. Some scientists believe they came to Australia from a northern land as long ago as

100,000 years, probably by means of boats. Before white men came to disturb their ways of life they were hunters or gatherers of food. They had no permanent homes, and not many belongings, but were very skillful.

The boomerang, a weapon that, in one form, circles back to its owner when it has been thrown, was invented by these people. From childhood they learned to hunt and track, and to find water when none could be seen anywhere on the dry plains. Rock engravings in New South Wales and other places, made by the aborigines, are considered to be among the finest examples of primitive art in the world.

The dingo, a type of dog, was probably brought to Australia by the aborigines. The animals thrived, and are now wild. As a group the dingos are outcasts and uncared for, and must find their own food by killing other native animals. Their future is a question.

For thousands, perhaps hundreds of thousands, of years, Australia was a haven. Men and animals lived at peace with each other, for there was room and food for all. Much of Africa was like this, too, a century ago. The Americas also were lands of promise and plenty until they were invaded by adventurers who were greedy for riches.

With the coming of the white man, all has been changed. As "civilization" advances, nature retreats. Forests are cut, hills and mountains are slashed and gouged, and swamps and wetlands, earth's great sponges and storehouses of water, are drained and filled. Litter covers the roadsides, garbage, raw sewage, and factory chemicals pollute the streams.

The dingo is a species of dog, scientifically called Canis dingo, *that is believed to have been introduced into Australia by the aborigines.*

Nature's Balance

Unless something is done, all nations on earth face the same bitter end. Nature is a giant network of living things, all in ever-changing but perfect balance. Man can upset the balance easily, and often does so, but correcting the mistakes he makes poses a problem he has not yet been able to solve.

Yet, all over the world, people are trying. Wildlife reserves and refuges are being set up where nature can restore herself, and the habits and needs of plants and animals alike can be studied. What scientists learn here can then be used in taking further steps for the full protection of living creatures, including human beings.

Each species, whether it seems to be useful to us or not, is the result of millions of years of development and change, and it has its own special place in the great web of life. Marsupials, and all other creatures, are of value. Some can be raised for food or for clothing. Others serve best as part of the framework of life, or add interest, enjoyment, or important new scientific information when they are used as subjects of research.

Human beings need food, space, and shelter. But, being human, they also need beautiful things to see, to hear, and to learn about, for the mind is as important as the body. We must have more than the bare necessities of life if we are to find peace within ourselves.

In this, marsupials and all other animals and plants have a place. And so do we.

Suggested Reading

Books

Australian News and Information Bureau, *Bush Dwellers of Australia* (Rigby Ltd., Adelaide, South Australia, 1967)

Beatty, Bill, *Unique to Australia* (Ure Smith, Sydney, New South Wales, 1962)

Breeden, Stanley and Kay, *Living Marsupials* (Winchester, New York, 1972)

———, *Wildlife of Eastern Australia* (Taplinger, New York, 1974)

Burton, Maurice, and Robert Burton, eds., *International Wildlife Encyclopedia* (20 volumes; Marshall Cavendish, New York, 1970)

Colbert, Edwin H., *Millions of Years Ago* (Crowell, New York, 1958)

Crowcroft, Peter, *Australian Marsupials* (McGraw-Hill, New York, 1972)

Darling, Louis, *Kangaroos and Other Animals With Pockets* (Morrow, New York, 1958)

Fenton, Carroll Lane, and Mildred Adams Fenton, *The Fossil Book* (Doubleday, Garden City, New York, 1958)

Grzimek, Bernhard, *Four-Legged Australians* (Hill and Wang, New York, 1967)

Hartman, Carl G., *Possums* (University of Texas Press, Austin, Texas, 1952)

Keefe, James F., *The World of the Opossum* (Lippincott, New York, 1967)

Kohn, Bernice, *Marvelous Mammals: Monotremes and Marsupials* (Prentice-Hall, Englewood Cliffs, New Jersey, 1964)

Lauber, Patricia, *The Surprising Kangaroos and Other Pouched Mammals* (Random House, New York, 1965)

Morcombe, Michael K., *Australian Marsupials and Other Native Mammals* (Scribner, New York, 1974)

———, *Wild Australia* (Taplinger, New York, 1972)

Petersen, Kai, *Prehistoric Life on Earth,* ed. and adapted by Georg Zappler (Dutton, New York, 1961)

Ride, W. D. L., *A Guide to the Native Mammals of Australia* (Oxford University Press, Melbourne, Victoria, 1970)

Schiller, Ronald, *Long Journey of the Continents,* in: Reader's Digest, *Marvels and Mysteries of the World Around Us* (Reader's Digest Association, Pleasantville, New York, 1972)

Serventy, Vincent, *The Singing Land: 22 Natural Environments of Australia from Surging Ocean to Arid Desert* (Scribner, New York, 1974)

Silverberg, Robert, *Forgotten By Time* (Crowell, New York, 1966)

Zappler, Lisbeth and Georg, *The World After the Dinosaurs* (Natural History Press, New York, 1970)

Magazines

Asimov, Isaac, "One-Animal Zoo" *(Science Digest,* July 1972)

Atkinson, Agnes Akin, "Br'er Possum, Hermit of the Lowlands" *(National Geographic,* March 1953)

Berry, James R., "What's Happening to Our World?" (*Science Digest,* June 1968)

Boardman, Walter S., "What Good Is An Opossum?" *(National Parks Magazine,* June 1964)

Breeden, Kay and Stanley, "Bashfulness Is Their Bag" *(International Wildlife,* Jan.-Feb. 1972)

———, "Eden in the Outback" *(National Geographic,* February 1973)

Colbert, Edwin H., "Antarctic Fossils and the Reconstruction of Gondwanaland" (*Natural History,* Jan. 1972)

Dickenson, Fred, "The Koala—Nature's Animated Toy" (*Reader's Digest,* Aug. 1968)

Dietz, Robert S. and John O. Holden, "The Breakup of Pangaea" (*Scientific American,* Oct. 1970)

Ealey, E. H. M., "Kangaroo Cave Life" (*Natural History,* Dec. 1963)

Fleay, David, "Flight of the Platypuses" (*National Geographic,* Oct. 1956)

———, "Strange Animals of the Island Continent" (*National Geographic,* Sept. 1963)

Frisch, Bruce H., "Our Moving Continents" (*Science Digest,* Jan. 1968)

George, Jean, "Supersaurus, the Biggest Brute Ever" (*Reader's Digest,* June 1973)

Griffith, Jeremy, "Search for the Tasmanian Tiger" (*Natural History,* Dec. 1972)

Haag, Wm. G., "The Bering Strait Land Bridge" (*Scientific American,* Jan. 1962)

Hallam, A., "Continental Drift and the Fossil Record" (*Scientific American,* Nov. 1972)

Hamilton, Wm. J., Jr., "Success Story of the Opossum" (*Natural History,* Feb. 1963)

Holmes, Charles H., "Australia's Patchwork Creature, the Platypus" (*National Geographic,* Aug. 1939)

Hurley, Patrick M., "The Confirmation of Continental Drift" (*Scientific American,* April 1968)

International Wildlife, "Member's Forum: Are Kangaroos the Underdogs Down Under?" (May 1974)

Johnson, David H., "The Incredible Kangaroo" (*National Geographic,* Oct. 1955)

Keefe, Jim, "Knee Deep in 'Possums" (*National Wildlife,* April-May 1967)

Kraft, Virginia, "Goodbye, Kangaroos" (*Sports Illustrated,* May 31, 1965)

Kurtén, Björn, "Continental Drift and Evolution" (*Scientific American,* March 1969)

Life, "Me, White Kangaroo" (April 26, 1963)

Matthews, Samuel W. and Wm. R. Curtsinger, "Antarctica's Nearer Side" (*National Geographic,* Nov. 1971)

Morcombe, M., "Swinging With Australia's Night Set" (*International Wildlife,* March 1974)

Nicholas, W. H., "Biggest Worm Farm Caters to Platypuses" (*National Geographic,* Feb. 1949)

Reader's Digest, "Nature's Little Puzzler: The Platypus" (August 1969)

Russell, F., "Death in Australia" (*Audubon,* March 1974)

———, "Marsupial Mysteries and Marvels" (*Audubon,* March 1974)

Schiller, George, "The Continents Are Adrift!" (*Reader's Digest,* April 1971)

Scholes, Wm., "Kangaroos, States, and Conservation" (*Science News,* June 6, 1970)

Science Digest,

"Antarctica's Most Distinguished Former Resident" (June 1968)

"East is Going West" (June 1967)

"Kangaroo Exports Prohibited" (April 1973)

"Little Package, Big Kick" (Oct. 1969)

"A Macropus by Any Other Name" (Oct. 1967)

"Meet the Ghost Kangaroo" (April 1964)

"Opossum: A Genuine Fraud" (March 1968)

"Tasmanian Devil" (Feb. 1971)

"Two for Extinction" (Feb. 1967)

"The Vanishing Bandicoot" (Dec. 1968)

Science News,

"Children Find Koala Ranges" (Dec. 9, 1967)

"Extinct Animals Found, Join Endangered List" (June 9, 1973)

" 'Extinct' Pygmy Possum Found Alive in Australia" (Oct. 22, 1966)

"Federal Controls Needed for Kangaroo Industry" (Oct. 1, 1966)

"Final Piece in the Gondwana Game" (July 27, 1974)

"Flying Start Made on Kangaroo Count" (July 31, 1965)

"Murder Bonus Paid on Wombats" (April 13, 1968)

"Protecting the Kangaroo" (Jan. 27, 1973)

"Rare Tiger Trace Seen" (Aug. 20, 1966)

"The Versatile Echidna" (Jan. 16, 1971)

Scientific American, "The Opossum as a Player" (December 1964)

Time, "Tie Me, Kangaroo, Down" (March 6, 1964)

Wadsworth, Nelson, "Colorado's 100-Foot Dinosaur: Is It the World's Largest?" (*Science Digest,* April 1973)

Index